RURAL SOCIO-ECONOMIC TRANSFORMATIOI
COMMUNICATION AND COMMUNITY DEVEL

PROCEEDINGS OF THE INTERNATIONAL CONFERENCE ON RURAL SOCIO-ECONOMIC TRANSFORMATION: AGRARIAN, ECOLOGY, COMMUNICATION AND COMMUNITY DEVELOPMENT PERSPECTIVES (RUSET 2018), BOGOR, WEST JAVA, INDONESIA, NOVEMBER 14-15, 2018

Rural Socio-Economic Transformation: Agrarian, Ecology, Communication and Community Development Perspectives

Editors

Rilus A. Kinseng, Arya Hadi Dharmawan, Djuara Lubis &
Annisa Utami Seminar

Department Socio-Economic Sciences, Bogor Agricultural University, Bogor, Indonesia

CRC Press is an imprint of the
Taylor & Francis Group, an **informa** business
A BALKEMA BOOK

CRC Press/Balkema is an imprint of the Taylor & Francis Group, an informa business

Typeset by Integra Software Services Pvt Ltd., Pondicherry, India

Published by: CRC Press/Balkema
Schipholweg 107C, 2316XC Leiden, The Netherlands

First issued in paperback 2023

ISBN: 978-1-03-257104-1 (pbk)
ISBN: 978-0-367-23603-8 (hbk)
ISBN: 978-0-429-28070-2 (ebk)

DOI: https://doi.org/10.1201/9780429280702

Publisher's Note
The publisher has gone to great lengths to ensure the quality of this reprint but points out that some imperfections in the original copies may be apparent.

Rural Socio-Economic Transformation – Kinseng et al. (Eds)
© 2019 Taylor & Francis Group, London, ISBN 978-0-367-23603-8

Table of contents

Community Development

Rural Socio-Economic Transformation – Kinseng et al. (Eds)
© 2019 Taylor & Francis Group, London, ISBN 978-0-367-23603-8

Foreword

The International Conference on Rural Socio-Economic Transformations on Agrarian, Ecology, Communication and Community Development Perspectives (RUSET 2018) was held in Bogor, Indonesia on November 14-15, 2018. The conference was organized by Department of Communication and Community Development Sciences, Faculty of Human Ecology, Bogor Agricultural University in collaboration with Asosiasi Program Studi Sosiologi Indonesia (APPSI/Association of Sociology Program Study in Indonesia). This conference seeks to answer questions such as how is the transformation of social, economic, cultural and ecological of the rural community taken place so far? What forces have and are at work that bring about rural transformation? What are the consequences of this transformation for the rural community and their environment? How far is sciences used in designing an intervention for rural transformation in order to make a better life for the rural community? To answer these questions we divided our theme into three topics: (1) agrarian and ecology; (2) communication and agricultural extensions; and (3) community development.

We would like to thank all authors for their efforts in preparing their papers. A great appreciation is also given to the reviewers for their assistance in reviewing the manuscripts. Special thanks to secretariat members for their assistance in formatting the layout of the proceedings.

Best regards,
Dr. Ir. Rilus A. Kinseng, MA.
Chairman

Rural Socio-Economic Transformation – Kinseng et al. (Eds)
© 2019 Taylor & Francis Group, London, ISBN 978-0-367-23603-8

Organizing committee

REVIEWERS

Rilus A. Kinseng
Department of Communication and Community Development Sciences, Faculty of Human Ecology, IPB University, Indonesia

Heiko Faust
Institute of Geography, Department of Human Geography, University of Georg-August-Universität Göttingen, Germany

Julie Ingram
Countryside and Community Research Institute, University of Gloucestershire, UK

Pudji Muljono
Department of Communication and Community Development Sciences, Faculty of Human Ecology, IPB University, Indonesia

Djuara Lubis
Department of Communication and Community Development Sciences, Faculty of Human Ecology, IPB University, Indonesia

Annisa Utami Seminar
Department of Communication and Community Development Sciences, Faculty of Human Ecology, IPB University, Indonesia

STEERING COMMITTEE

Dr. Ir. Arya Hadi Dharmawan
Prof. Endriatmo Soetarto
Dr. Ir. Nurmala K. Pandjaitan
Prof. Sumardjo

ORGANIZING COMMITTEE

Dr. Rilus A. Kinseng
Bayu Eka Yulian, MSi
Rizka Amalia, MSi

WORKING GROUP 1. AGRARIAN AND ECOLOGY

Dr. Soeryo Adiwibowo
Dr. Ekawati Sri Wahyuni

Dina Nurdinawati, MSi
Raisita, MSi

WORKING GROUP 2. COMMUNICATION AND AGRICULTURAL EXTENSION

Dr. Djuara Lubis
Dr. Sarwititi Sarwoprasodjo
Titania Aulia, MSi
Dr. Annisa Utami, MSi

WORKING GROUP 3. COMMUNITY DEVELOPMENT

Dr. Ratri Virianita
Rajib Gandi, MSi
Ade Mirza Roslinawati, SKPm
Widya Hasian Situmeang, SKPm

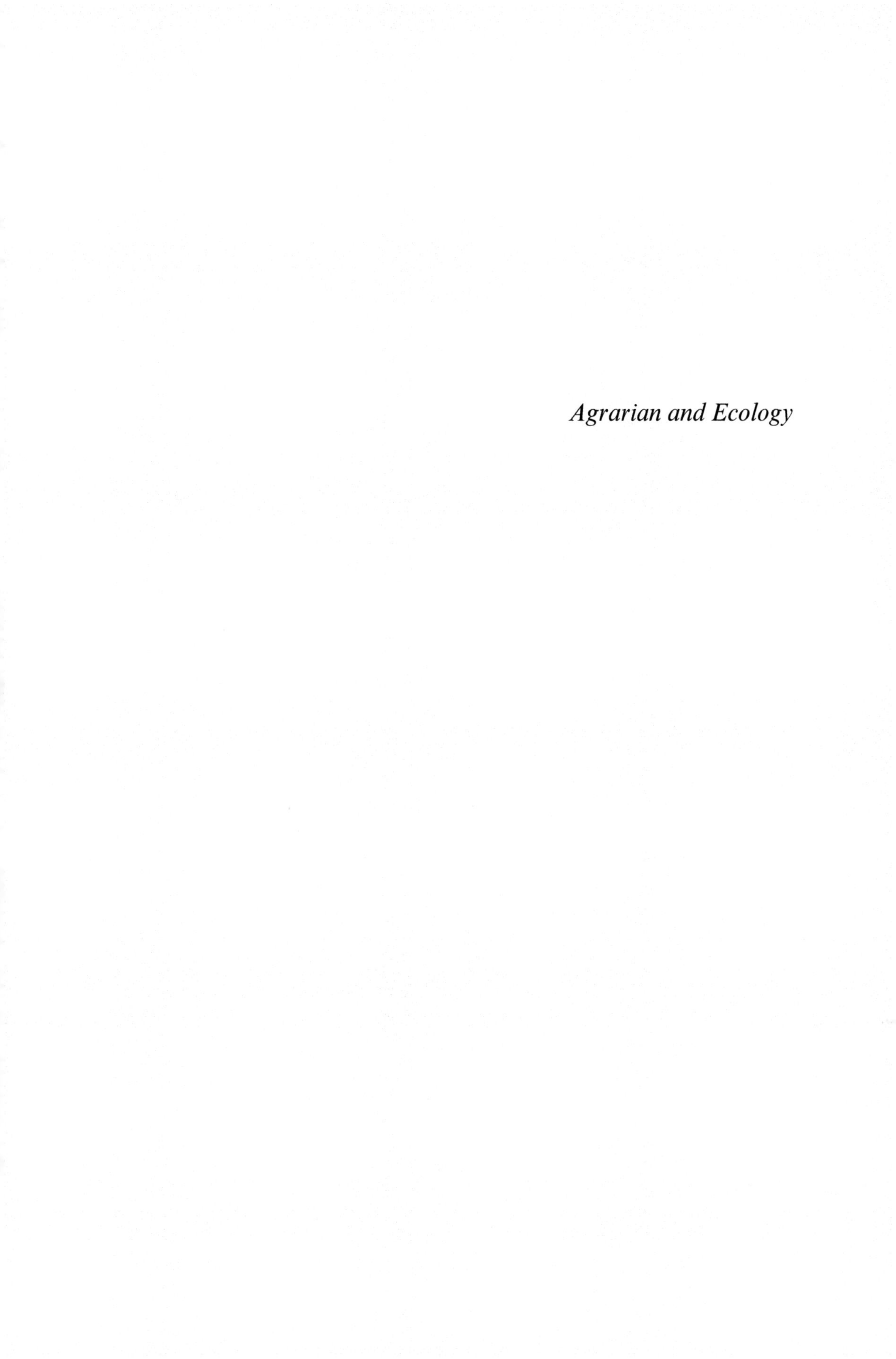

Agrarian and Ecology

Rural Socio-Economic Transformation – Kinseng et al. (Eds)
© 2019 Taylor & Francis Group, London, ISBN 978-0-367-23603-8

The typologies and the sustainability in oil palm plantation controlled by independent smallholders in Central Kalimantan

A. Andrianto
Center International Forestry Research, Bogor, West Java, Indonesia

A. Fauzi & A.F. Falatehan
Bogor Agricultural University, Bogor, West Java, Indonesia

ABSTRACT: In 2016, Indonesia's oil palm plantations reached an area of 12.3 million ha, of which 38.6% or 4.8 million ha were smallholder plantations and approximately three-quarter were identified as independent smallholders. Independent smallholders are the most unproductive oil palm producer group in Indonesia experiencing high compliance barriers caused by different abilities to adopt good agricultural practices, access to input produc-tion and formalize operations. Using Central Kalimantan as a case study, research intend-ed to contribute to efforts to unpack the diversity of independents oil palm farmers in In-donesian. We found 6 different typologies obtained by cluster analysis. Further analysis shown each typology has different sustainability index. Sustainability performance is closely related to knowledge and experience thus influence the ability and motivation to implement and conform to the sustainability standard. Our findings suggest focusing on the sensitivity of each typology to perform selective upgrade and simultaneous control of the smallholder expansion.

1 INTRODUCTIONS

In 2016 oil palm plantations in Indonesia reached 12.3 million ha, of which 38.6% or 4.8 million ha were smallholder plantations (DJP 2017) and three-fourth of them are identified as inde-pendent smallholders (Jelsma et al. 2017, INOBU 2016). Independent smallholders (IS) are a very fast-growing group both in terms of the number of farmers and the area planted. However, IS are increasingly marginalized, in contrast to the plasma farmers. IS are the most unproductive producer group (Molenaar et al. 2013, Euler et al. 2016) and experiencing very high compliance barriers caused by different abilities to adopt good agricultural practices, access input production and formalize operations (Prokopy & Floress 2008, Brandi et al. 2015). Those constraints will limit their market access (Brandi et al. 2015, Rietberg & Slingerland 2016). Moreover, the 6 large scale private oil palm companies have made additional commitments to eliminate deforestation from their supply chains, which significantly improved their monitoring and traceability. Con-cerns began to increase over the threat of corporate consolidation towards their sources of sup-ply, consequently the alienation from formal market and market bifurcation risks (Gnych et al. 2015, Jopke & Schoneveld 2018).

Different abilities as revealed above occur because IS are very diverse producers, as shown in Sumatra (McCarthy & Zen 2016, Jelsma et al. 2017). There were tremendous differentiations re-garding the resources, capabilities, livelihood strategies and relations with markets. The negative externalities for the environment, according to Jelsma et al. (2017) and Potter (2016), referred to how many "entrepreneurial farmers", in the context of increasing land scarcity, then converting fragile forest ecosystems and establishing oil palm plantation.

This paper is based on CIFOR exploratory survey research in Kalimantan with main focus to contribute in unpacking the diversity of IS. The first section presents IS typology obtained by cluster anaysis. Secondly, we analyzed each typology using composite analysis to determine

Table 1. Smallholder mapping results and sampling.

District	Area with Independent Small holder	Sample (ha)	Number of grids sampled	Number of Survey		
				Rapid plot	Farm	Farmer
West Kotawaringin	42,700	2,902	195	767	349	308
Pulang Pisau	2,763	737	113	231	150	138

Source: CIFOR (2016).

sus-tainability level. The results findings then are discussed within main sustainability issues and how to solve the problems on each typology.

2 METHOD

2.1 *Site selection and survey activities*

Research were conducted in two smallholders with different context of ecological and economic landscapes in Central Kalimantan, Indonesia. We used CIFOR baseline map of IS where plots were randomly sampled by cells 500m*500m. The result of mapping survey activities are presented in Table 1.

The rapid plot survey obtained 998 IS plots in the sample area. At each selected plot, we conducted farm and farmer surveys. The farm survey involved observation data, topography, soil type, quality of on-farm infrastructure, planting patterns, weeding practices, presence of rotten Fresh Fruit Bunches (FFBs), cover crops, erosion and canopy cover. At the middle of the plantation, 20 trees were sampled, with five tree intervals, to determine the presence of pest and disease, nutritional deficiencies (specifically P, K, Mg and B), quality of pruning, and palm varieties at a tree level. The methods based on the field audit standard developed by Fairhurst & Griffiths (2014). A farmer questionnaire was subsequently administered with the plot's owner. The questionnaire comprises of questions on household characteristics, household livelihood activities, types of oil palm plantation owned, production practices, labor allocation to oil palm, nature of linkages to input and offtake markets, participation in farmers groups, legality and household strategies.

2.2 *Typology analysis*

The aim of cluster analysis, generally, is to differentiate between the objects. There are different ways to explore heterogeneity amongst smallholders and to construct a smallholder typology (Alvarez et al. 2018). The variables used for the clustering are listed in Table 2. A two-step cluster analysis by SPSS software was chosen because it can combine numerical and categorical data. Finally, we profiled the cluster solution using the observable variables.

2.3 *Sustainability analysis*

The term of 'sustainable' define when the development met the needs of the present without compromising the future generations to meet their own needs (Brundtland 1987). In the practical level, as a socio-ecological process, it is characterized by the pursuit of a common ideal - unattainable in a given time and space (Wandemberg 2015). Sustainability is assessed and commonly defined as a tool to identify, predict and evaluate potential environment, social and economic impacts of an initiative to assess sustainability, which is a weak sustainability (Diesendorf 2000).

In this study we used variables (Table 3) that are of major concern in the field of environmental economics and natural resources: property right objectives, ecology objective, management objectives, and economic objectives (Hussen 2004, Tietenberg 2009). Those variables are implicitly accommodated in: (1) Indonesian Sustainable Palm Oil (ISPO) state in Agricultural

Table 2. Variables used for cluster analyses.

Variables	Sub-variables
Natural resources asset	1. The plot size of oil palm plantations 2. Year of land acquisition 3. Year of starting oil palm planting 4. Land status 5. Type of soil
Social capital	1. Farmer education level 2. Oil pam farm management experience 3. Membership of farmers group 4. The first party to give/sell the land 5. Ethnicity 6. Originate from 7. Recent residential
Finance	1. Source of initial investment in developing palm plantations 2. Prior income before planting oil palm 3. Main current income 4. Revenue contribution from oil palm
Labour	1. Household workers involved in 2. External workers recruited individually 3. External workers recruited collectively

Table 3. Variables used for cluster analyses.

Variables (Key Performance Indicator) KPI	Measurements
1.1.1 Land ownership status	The order of the legality degree of land ownership titles are: SHM certificate, notary deed, girik, seal paper, SKT, and no letter
1.1.2 Cultivation registration/STBD	Already registered = 10, in process 8–9, the docs for registration is incomplete 7. The lowest value of 0 is not knowing
1.2.1 Location of oil palm plantatation with forest zones	Overlay the plots with forest map (APL, HPK HP, HPT, HK). Score 10 located on APL (non-forest) zone, the lowest on conservation area
1.2.2 Location of oil palm plantations with peat protected areas	Overlay the plots by peat protection maps. If plot outside the peat protection area on mineral soil = 10, outside the peat protection area but containing peat scores: 6–9. Plots inside peat protection area 5, the deeper the peat the smaller the value.
1.3.1 Recognition from community/ land conflict	Recognition means no conflict with community, companies, Government: 10 points, if there is a conflict with 1 stakeholder, will decrease one point. The lowest value if the conflict becomes anarchic
1.3.2 The level of activity in farmers group	Max values are active members in farmers group and enjoy benefits, the lowest value is not farmers group because they don't want to
2.1.1 Productivity of oil palm plantations	Palm productivity at a certain age compared to the ideal standard of PPKS. If the production is the same as the PPKS standard = 10. Every reduction of 500 kg ha/year, will reduction 1 point. The lowest if the differences more than 5 ton compared to the standard
2.2.1 FFB sales increases 1	The fewer chains the better. Direct sales to the factory = 10. If transaction chain the value decreases by 2 points.
2.2.2 FFB pricing	

(*Continued*)

Table 3. (*Continued*)

Variables (Key Performance Indicator) KPI	Measurements
	Max value if the price matches with Government's benchmark and the lowest if the differences are wide and no room for negotiation
2.2.3 Number of buyers	The number available buyers, the more the higher the value.
3.1.1 Seed source provider	The highest score is certified seed from certified local agents, then company assistance, government assistance, buying directly from the company, seedlings from natural tillers, the lowest is not knowing the source
3.1.2 Palm oil variety	Assessed the proportion of the 20 palms. The max score if all are tenera. The lowest if all palms are dura variety.
3.2.1 Cropping pattern	The ideal spacing between each plant is around 9 m, will get a score of 10. The score will be reduced if it does not match the ideal spacing and irregular cropping patterns
3.2.2 Harvesting path conditions	Max the path is well maintained and easy to pass.
3.2.3 Dead leaf	Assessed the proportion of the 20 oil palm trees surveyed in the plot. Max
3.2.4 Circle weeding	Max value if all are in accordance with GAP. 20 trees are fit = 10,
3.2.5 Rotten oil palm fruit	if 2 trees do not match = 9, if 4 trees do not match = 8, and so on
3.2.6 Collecting of palm fruit loose	
3.2.7 Pruning	
3.2.8 Oil palm health condition	
4.1.1 Initial land cover berfore oil palm plantation	Max score, if farms were built on bare land/not productive and the lowest value if converting peat primary forest for oil palm plantation.
4.2.1 Water conservation	Max if existing canal is blocked, The lowest if no canal blocks.
4.2.2 Flood experience	Max score if no flooding experienced and the lowest if recurring floods occurred in recent years.
4.3.1 Soil conservation	Max score if planted cover crop and made mounds in the farm, the lowest value if the soil was left open entirely.
4.3.2 Deficiency Potassium (P)	By indicators of the oil palm condition. Assessment is given
4.3.3 Deficiency Kalium (K)	through the degree of deficiency in 20 plants surveyed. Max value =
4.3.4 Deficiency Magnesium (Mg)	10 if no indication of deficiency. The more deficiency indications
4.3.5 Deficiency Boron (Br)	the less the score. The lowest value if plants experience nutrient deficiencies at a serious level.
4.4.1 Experience fire and fire use	Max score if not used fire and never been a fire. The lowest value if using fire and recurring fires.

Ministry of the Republic of Indonesia No 11/2015, (2) good agricultural practices (DJP 2014) by the Directorate General of Plantations, Ministry Agriculture, and (3) principles and criteria for sustainable palm oil production by RSPO (Roundtable on Sustainable Palm Oil).

The sustainable smallholders palm oil assessment framework procedure follows as is shown in Figure 1 (Lim 2015). Whereas for overall sustainability index assessment might be constructed from the composite objective data (Fauzi 2014, Lee 2014). Each objective consists of numbers of Headline Performance Indicators (HPI). Each HPI consists of numbers of Key Performance Indicators (KPI). KPI describes as HPI's key impact areas of each respect to palm oil production that can foster or impede the achievement of each sustainability objective. The KPI score is obtained from the analysis of field survey data on each variable. KPI assessed using a Likert scale of 1–10, depending on the performance under the conditions of environmental, economic, oil palm management and the property right.

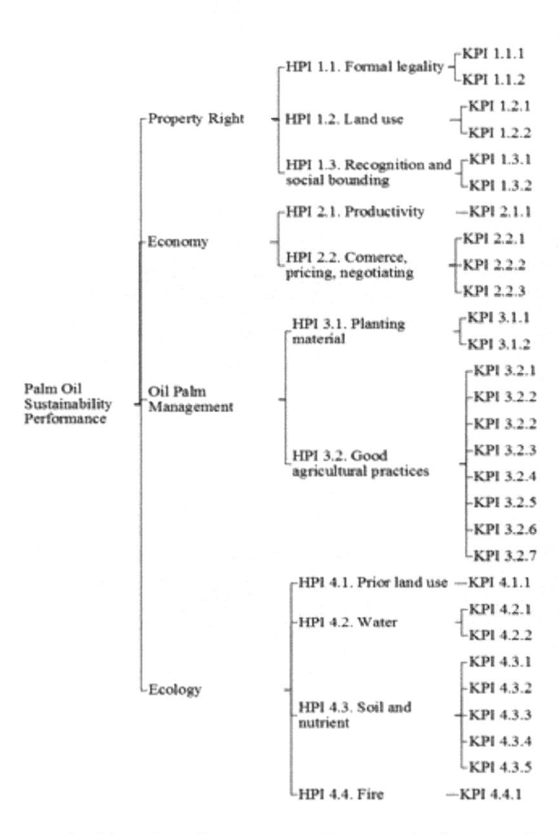

Figure 1. Smallholders sustainable palm oil assessment framework. Source: adapted from Lim (2015).

3 RESULT: TYPOLOGY AND SUSTAINABILITY

3.1 *Smallholders typologies*

From the rapid survey we identified 998 oil palm plantation plots, and 499 plots (owned by 446 farmers) were chosen for farmer and farm surveys. We dropped 10 farmers' data due to incomplete and missing data. The final analyzed were 489 plots (owned by 436 farmers). Using 19 distinguishing variables, we classified the group into 6 typologies (Table 4).

3.1.1 *Typology 1: migrant laborers*

Settlement and plots of palm oil plantation were located within the same sub-district. In general, they were migrants from outside Kalimantan, either under transmigration program or by spontaneous migration that were employed by the oil palm companies. With knowledge and experiences obtained during the employment, they developed their own oil palm on their

Table 4. Typologies of smallholders' oil palm.

Cluster	N	% of total	Profile cluster
1	77	17.7	Migrant laborers
2	85	19.5	Early adopters
3	56	12.8	Entrepreneur
4	68	15.6	Elites
5	51	11.7	Subsistence farmers
6	99	22.7	Migrant farmers
Total	436	100	

lands. The group could possess between 3 and 15 ha by purchasing bare lands using their own savings from years of employment at the oil palm company. The lands were mainly located on mineral soil, with 10% of peat deposits.

3.1.2 *Typology 2: early adopters*

Majority of the group were native to Dayak and Malay, including minor second generation of migrant families. The group were pioneer farmers in the locations where they reside and grow fields. The farms were the oldest in the neighborhood. They managed nearly all daily works by themselves. In average, they could manage between 3 and 15 ha of land, which obtained either by purchasing, by familial inheritance, or from the government. Roughly about 40% of the land had certificate of ownership (SHM). A half of the total farms were converting from forestlands and agricultural lands on peat lands.

3.1.3 *Typology 3: entrepreneurs*

The group consists of multi ethnics both native originally and migrants. The domiciles spread widely within the village, outside the sub district, outside the district, and even outside the province. Many farms, which were belong to the group, had been mandated to local managers. Each type of field work is carried out by paid labor. Although the land owners were educated people, they have a little experience upon starting investment on oil palm business. The major incomes sourced from dealing business and running oil palm fields above 50 ha. They hold, more than 50%, SHM land certificate. Half of the oil palm farms of the group are grown on bare lands and the rest 50% are of forest conversions, including peat lands.

3.1.4 *Typology 4: elites*

Each owner within the group has 15–50 ha. They were influential individuals with strong connections to the plots where the farms locate. They comprised of local people and long residence migrants. In fact, they did not belong to farmers group, but the 20% of them take part in the farmer group as the leader. The group were known to be highly educated with main occupation as civil government, shop owners, and palm oil bosses. 70% of the land in the group owned by purchasing and only 20% were obtained from the government, with land legality in the form of SKT.

3.1.5 *Typology 5: subsistence farmers*

The group possessed less than 3 ha, only 30% of the group owned 3-15 ha. One-third of the typology owned the land from familial inheritance and the government. Three-fourth of the plots were converted from rubber plantation and agriculture lands. The rest of 25% was built on empty land. Some of the group belong to farmers group and other were individually managed the plots. Either during non-harvest period or working the land, they seek for additional income by laboring for daily wages.

3.1.6 *Typology 6: migrant farmers*

Migrant farmers were migrants from outside the province either under transmigration program or by spontaneous migration. Most of the group resided in the same village where the plots were located. However, only 10% belongs to farmers groups. They mostly conducted farm field activities by themselves. 60% of the group had plot around 3–15 ha, the remaining had less than 3 ha. They planted oil palm on the bare land and the forested land. In contrasts to the subsistence farmers group, migrant farmers were expanding their farm by planting rubber for livelihood strategies.

3.2 *Smallholders sustainability index*

As a result of typology analysis, we then analysed the performance of each sustainable variables (Table 3) based on equations (1), (2) and (3) in each typology. The results of the analysis are shown in Table 5.

Table 5. Overall sustainability index assessment for each smallholders' typologies.

KPI	Score for KPI						Score for HPI						Sustainability value	
	C1	C2	C3	C4	C5	C6	C1	C2	C3	C4	C5	C6	Objective	Over all
1.1							4.6	3.7	7.0	2.2	4.1	3.6	C1: 6.3	C1: 5.8
1.1.1	4.9	5.7	7.7	3.1	5.7	6.3							C2: 6.5	C2: 5.5
1.1.2	4.3	1.6	6.3	1.3	2.5	1.0							C3: 7.4	C3: 6.2
1.2							8.7	8.5	9.0	8.6	9.2	9.4	C4: 6.4	C4: 5.0
1.2.1	8.8	8.6	8.9	9.2	10.0	9.7							C5: 7.5	C5: 5.3
1.2.2	8.7	8.4	9.2	8.1	8.4	9.2							C6: 6.7	C6: 5.6
1.3							5.5	7.3	6.2	8.1	9.1	7.1		
1.3.1	3.4	9.5	9.3	10.0	10.0	9.9								
1.3.2	7.5	5.1	3.0	6.2	8.2	4.2								
2.1							1.1	2.1	5.2	2.0	1.2	1.2	C1: 2.8	
2.1.1	1.1	2.1	5.2	2.2	1.2	1.2							C2: 3.2	
2.2							4.4	4.3	4.2	3.7	3.7	3.9	C3: 4.7	
2.2.1	5.3	6.2	7.7	5.1	5.0	5.7							C4: 2.9	
2.2.2	2.1	2.1	2.0	2.0	2.2	2.1							C5: 2.5	
2.2.3	5.9	4.5	4.9	3.9	4.0	4.0							C6: 2.6	
3.1							5.8	4.4	5.8	4.0	4.4	5.1	C1: 6.9	
3.1.1	8.8	5.5	7.0	5.9	6.3	6.1							C2: 5.4	
3.1.2	2.8	3.4	4.6	2.2	2.6	4.2							C3: 6.4	
3.2							7.9	6.3	6.9	5.7	5.2	7.1	C4: 4.9	
3.2.1	9.7	5.6	7.5	8.2	3.1	4.8							C5: 4.8	
3.2.2	9.9	5.2	7.7	7.9	7.5	7.8							C6: 6.1	
3.2.3	9.9	5.2	7.7	7.9	7.5	7.8								
3.2.4	9.5	7.8	8.4	4.6	5.5	9.4								
3.2.5	2.2	5.9	4.5	1.8	2.7	7.5								
3.2.6	7.3	4.3	4.5	2.3	3.3	3.3								
3.2.7	9.8	9.4	8.9	8.4	9.6	9.3								
4.1							9.3	7.2	6.8	6.5	7.6	5.5	C1: 7.2	
4.1.1	9.3	7.2	6.8	6.5	7.6	5.5							C2: 6.9	
4.2							7.9	7.7	6.6	6.1	6.8	8.4	C3: 6.2	
4.2.1	7.7	7.0	5.9	5.8	5.8	8.1							C4: 5.9	
4.2.2	8.2	8.4	7.3	6.3	7.8	8.6							C5: 6.3	
4.3							6.0	5.4	5.0	5.0	4.8	5.4	C6: 7.0	
4.3.1	6.6	5.2	5.2	6.3	4.5	5.5								
4.3.2	7.3	7.0	6.3	5.5	6.1	6.2								
4.3.3	2.0	1.6	0.8	1.5	1.2	1.0								
4.3.4	7.0	6.9	6.3	5.5	5.2	7.7								
4.3.5	6.9	6.1	6.3	6.1	7.1	6.5								
4.4							5.8	7.3	6.3	6.0	5.9	8.7		
4.4.1	5.8	7.3	6.3	6	5.9	8.7								

* KPI score was calculated by the average of key performance value of all cluster members.
* C1 = Migrant laborers, C2 = Early adopters, C3 = Entrepreneurs, C4 = Elites, C5 = Subsistence farmers, C6 = Migrant farmers.

4 DISCUSSIONS

4.1 *Independent smallholder typologies*

The six typologies of IS are successfully distinguished based on 19 sub-variables grouped into 4 variables as collected and analyzed through farm survey and household surveys, using random sampling method on oil palm plantation plots in two districts (Table 1). Naming the terms of the six typologies is developed from variables that associate with each other. The identity of typologies refers to various variables that are associated between one another. It was mentioned as an association because all of them are the re-sults of interactions of various extreme variables that appear and/or do not appear in each typology. The survey method in this study relying on such questions as who the local peo-ple are, the actual size of plots owned by IS and the absentee plots, was successful in more accurately clarifying issues, often debated, missed or often not explored fully in many surveys. Several studies using large sample household surveys and targeting solely farmers living in the village (Molenaar 2013, Lee 2013, Daemeter Consulting 2015, Euler 2016 and Winrock & SPKS 2018) seem to fail to address properly what local people or IS are meant, who are the ethnic, how long particular smallholders have been living in partic-ular place.

Compared to surrounding people who did not adopt oil palm plantations, in general oil palm planters are categorized as good economic communities. Most oil palm farmers had oil palm plots of 3–15 ha, while those that had under 3 ha only 17.27%, mostly in the sub-sistent farmer typology and less than one third in the migrant farmer typology. Groups that have oil palm plots between 15–50 ha are dominated by elite typology and those with more than 50 ha are dominated by typology entrepreneurs. However, even though the group that owns more than 50 ha of land, only a few are in the group of entrepreneurs and some elites, but they control more than half of the land of independent smallholder oil palm plantations.

Farmers in various typologies employs social capital resources from the very beginning when they had idea of planting oil palm, establish and manages the plantations, to the stage when they harvest and secure the yields from their farm land. The facts are in line with the opinion of Baker (1990) which states that resource actors derive from specific social structures and then use their own interests; it is made by changes in the relationships among actors. Migrant laborers utilized their similar bond of fate as migrant workers on overseas lands. They shared information of available land for oil palm and would super-vise plantation management. Early adopter used in social capital as old people and pio-neer, knew the history of land. Entrepreneurs who normally lack strong social capital ties with local communities, would minimize conflict by establishing social capital bonds with community and local leaders by providing assistance and giving them loans to make latter become dependent. As local figures, Elite typology, resident or migrants have important positions in the government. They use the power of formal figures and informal figures at-tached to them to acquire and secure land. The people are happy with the elite existence as a protective figure. Subsistent farmers, as a marginal village community, establish farmers group to strengthen the collectivist position in securing land and farm management. Mi-grant farmers share knowledge informally to support the identity as serious migrants in managing both agricultural and oil palm.

Nearly all IS are supported by personal financial resources and loan assistance from families. But in recent years there has been an increasing utilization of bank credit ser-vices. There has been more than 3% of the samples known to have utilized bank services. They have access to bank credits with guaranteed oil palm plantations that are already producing. However, for self-help oil palm plantations, this financial scheme is rather dif-ficult to determine whether they are using bank credits fully or not since the financial management of the households, both income from various sources and expenditures for various needs, are still made mixed, not segregated.

While it is generally assumed that smallholders work on self-managed by owner's fami-ly, our research reveals different results. Almost each typology employs hired external la-bor services and many of family members are not involved in. Even though we were suc-cessful in identifying domestic workers and hired labors, we could not unfortunately seg-regate labor by each working days due to large data gap regarding the outpouring of labor. Factors that influence the owners of the plantation to get involved in management are: mo-tivation, livelihood choice, expertise, time availability, the need to control high-cost work, and production. Those explained why 2/3 of migrant laborers involved in fertilization, half of early adopters manage their own farms mainly in harvesting, 30% of entrepreneurs only supervise, elites entrust management to field managers, subsistence farmers can only do half of all field work, migrant farmers involved in the overall work activity.

4.2 *Smallholders typologies and sustainability*

The wide gap variation in the sustainability index amongst smallholders' typologies are quite clearly seen on a scale of 1 to 10. If sorted from the lowest are the elites (5.0), subsistence farmers (5.3), early adopters (5.5), migrant farmers (5,6), migrant laborers (5,8) and entrepreneurs (6,2). However, the differences in index occur because in dealing with the same problems each typology responds differently depending on their capacities to find solutions.

The toughest problem independents smallholders' face in this HPI is formal legality of the land. Even though, entrepreneurial group seem to have pretty good value, but they are only able to "conditioning" and complete the two KPIs at the district level which are inadequate. They tend to own plots whose sizes are is relatively large and exceeds the limit al-lowed by regulation (max 25 ha). Those owning land of more than 25 ha are required to get an HGU in order to clarify the land control and to get IUP permit to validate cultivation business. For the HPI Land Use, in general the values of each cluster are relatively good. But when analyzed more deeply about the value that reduces HPI occurs in relatively comes from new oil palm plots. In other words, the development of smallholders' plantation expansion will have been increasingly penetrated forest areas and peat areas. On HPI recognition and social bounding, the most difficult problem faced by the typology of mi-grant laborers is land conflicts. This is because they got land from intermediaries, where much of the land turns out to be not in accordance with the boundaries of the land offered so they must pay extra as a loss to the owners of the land whose land they use. For the KPI farmers group in this HPI, most each cluster members face obstacles because their plots located outside the village they live.

The way how migrant farmers, local elites, and entrepreneurs are performing in terms of ecology is indicated by the trajectory of the lands before they established plantations. Due to limited capital, there is a tendency among migrant farmers to invest in planting oil palm but at a low cost, making them to look for cheap and readily, and often forested lands. Similar to this group, elites and entrepreneurs aside from considering low prices, they also want a large plot on one stretch. Most of areas where they clear to pave way for plantation are still forested, which is potentially causing the loss of biodiversity and environmental services. In addition, the relatively large portion of the lands controlled by entrepreneurs, elites, and subsistence farmer groups is peat. Their common practice of planting palm oil is preceded by drying the lands and the peat, as many of their plots are partially flooded. They caused canals to become dried and could unfortunately not afford to build canal blocking to maintain the minimum water level and make the peat remained moist. During dry season, such dried peatland are subject to fire. This poses a challenge for the three groups to perform in terms of how they maintain water resources as specified under HPI.

In general, farmers are less concerned about maintaining soil and nutrient content. Lack of awareness and knowledge among most of the IS groups about the importance of soil conservation cause them to be less skillful in building mounds and planting cover crops. Our survey indicates that K deficits are high in the plots of each cluster, which give a signal that famers pay less attention to maintaining the nutrient content. Potassium is the most absorbed element by oil palm, and it lost due to easy washing and evaporating. Although the value of HPI fire in each cluster is above 5.8, the value of this indicator is an important aspect because it has a large and broad impact. We found that farmers in each cluster still uses fire to clear land and burn waste on the land. So, it is not surprising to find that the plots affected by the fire are plots where farmers are still using fire to clear the land.

Planting material constitutes the biggest common problem for smallholders, as indicated by the lowest KPI score on oil palm varieties, which range from 2.2 to 4.6. We found a high percentage of duratype plants on plots in each typology. Entrepreneurs' typology have relatively high HPI score because they have direct access to seed providers from the PPKS Medan and know how to access seeds from oil palm plantations companies near their plots. However, we found an anomaly in the typology of migrant laborers where they stated that they obtained seeds from certified local seed suppliers. But it turns out that the types of varieties found after harvest are not appropriate. In this case, it is also necessary to supervise local palm oil breeders, so as not to mix certified seeds with seeds of un-known origin. On HPI good agricultural practices the common problems that occur are lack of proper pruning, not putting the leaf midribs dead properly, and not collecting palm fruit loose fruit until nothing left.

Economy objectives are the biggest problems faced by smallholders. The problem with HPI productivity is very closely related to the objectives of ecology and objectives of oil palm management. The scores on this HPI is between 1.1 and 5.2. The productivities gaps are very wide when compared to the standard PPKS production, presumably due to lack of use appropriate fertilizer, the right dosage and time specified. Unfortunately, this study is not able to confirm because much of the data obtained from fertilization is incomplete, because farmers do not have records and are unable to remember well, except in the typology entrepreneurs. Entrepreneurs have a relatively high productivity value for small-holders because they use consultant services to determine fertilizer needs and could obtain needed fertilizers from outside the province. The same thing happened to HPI commerce and price. Considering the variation in KPI, the entrepreneurs' typology positioned itself at the best level, because they made sales directly to the mill. While the other typologies are very dependent on intermediary buyers. Although the number of buyers are quite large, the prices are lower than the benchmark price and there is no room for negotiating the price.

4.3 *Towards sustainable smallholders*

By recognizing obstacles in different typologies, efforts can be made to selectively upgrade farmers to help them adopt sustainable practices. Upgrading typology migrant laborers on the property right objective can be done by providing facilities and assistance to obtain STDB and land certification. In the Ecological Objective, soil conservation needs to be done through fertilizing and planting cover. Increasing the management objective seems that they only need to survive because the lowest value is due to plant varieties but productivity can be increased by training and at the same time providing the right fertilizer. For the economy: increasing productivity and direct access to factories or collectively selling.

To upgrade the early adopters typology, attention can be paid to improving the level of productivity. While their plantation is still young and has not reached the maturity or old level, there is a possibility to include them as a recipient group of the Ministry of Agriculture's replanting programs funded by palm oil fund. Their low productivity of crop make them eligible to receive assistance and funding. This should also be accompanied by strengthening the farmer institutional and their farm technical capabilities.

From the aspect of property right, primarily land legality, the typology entrepreneurs need special treatment. They need to be encouraged to obtain an IUP through applicable procedures. From ecological aspects, the group also needs special attention in terms of how they are using fire and conserving water and maintaining hydrology by making canal blocks around the plots. Special attention should also be paid to the elites' typology because many of their plots are located within the forested areas and peat protection area. A special mechanism needs to be done wisely to maintain this field for one cycle through the TORA (Tanah Obyek Reforma Agraria – land object for agrarian reform) or PS (Perhutanan Sosial - social forestry) program, and perhaps some of the plantation fields need to be restored to return back to its previous function.

The typologies of subsistence farmers and migrant farmers are relatively similar. They are both traditional planters and hard worker with limited capital capabilities. To improve their capacity in managing oil palm, these two groups need to be facilited in obtaining "smallholders oil palm registration (STDB)" document and getting their lands legalized through land certificate. In addition, institutional counselling and oil palm cultivation training need to be provided so that they could increase technical capability in handling most of the field activities. This includes providing credit capital assistance for purchasing fertilizers.

5 CONCLUSION

Survey used area-based sampling (not farmers-based sampling) with a large enough number of samples; which objectively revealed the diversity of the oil palm plantation and its farmers.

This method could explain the variety of independent smallholders oil palm and eliminate research bias samples because of the subjectivities of the researchers.

The uniformity grouped in each typology indicates the characters of the farmers and their ability to manage the plots. These typologies are in line with the efforts, motivations, and constraints of the farmers. Upon careful assessment, we found the typology has a strong relationship to every objective of sustainability. Sustainability performance is closely related to knowledge and experience; as well as the ability and motivation to implement.

Accordingly, efforts to resolve sustainability issues should not focus on technical aspects, bureaucratic procedures, and administration in the plantation sector only; but also, on the governance aspect outside the plantation sector. Learning from the success of each typology, we proposed to apply the same method in finding solutions to problem solving; to improve the governance of the plantation sector. The combines variables of each typology can be used as an entry point to closer the sustainability gap, namely using the power of existing social capital so that governance outside the plantation sector (finance, labor, industry, forestry) related to the plantation sector will help accelerate solving the problem of independent oil palm farmers simultaneously.

REFERENCES

Alvarez, S.T., Michalscheck, C. J., Paas, M., Descheemaeker, K. Tittonell, P. & Groot, J.C. 2018. Capturing farm diversity with hypothesis-based typologies: an innovative methodological framework for farming system typology development. *Plos one* 13(5), e0194757.

Brandi, C., Cabani, T., Hosang, C. Schirmbeck, S. Westermann, L. & Wiese, H. 2015. Sustainability standards for palm oil: challenges for smallholder certification under the RSPO. *Journal of Environment & Development* 24(3): 292–314.

Brundtland, G.H. 1987. *Our Common Future. World Commission on Environment and Development (WCED)*. Oxford, UK: Oxford University Press.

Carlson, K.M., Curran, L.M., Ratnasari, D., Pittman, A.M., Soares-Filho, B.S., Asner, G.P. & Rodrigues, H.O. 2012. Committed carbon emissions, deforestation, and community land conversion from oil palm plantation expansion in West Kalimantan, Indonesia. *Proceedings of the National Academy of Sciences* 109(19): 7559–7564.

Casson, A. 2000. *The hesitant boom: Indonesia's oil palm subsector in an era of economic crisis and political change*. Bogor: CIFOR.

Daemeter Consulting. 2015. *Indonesian Oil Palm Smallholder Farmers: A Typology of Organizational Models, Needs, and Investment Opportunities*. Bogor: Daemeter Consulting.

Direktorat Jenderal Perkebunan (DJP). 2017. *Statistik Perkebunan Indonesia Kelapa Sawit: 2015–2017*. Jakarta: DJP.

Direktorat Jenderal Perkebunan (DJP). 2014. *Pedoman Budidaya Kelapa Sawit Yang Baik*. Jakarta: DJP.

Diesendorf, M. 2000. Sustainability and sustainable development. In Dunphy, D. Benveniste, J. Griffiths, A. Sutton P (eds). *Sustainability: The Corporate Challenge of the 21st Century* 2:19–37. Allen & Unwin. Sydney.

Euler, M., Hoffmann, M.P., Fathoni, Z. & Schwarze, S. 2016. Exploring yield gaps in smallholder oil palm production systems in eastern Sumatra, Indonesia. *Agricultural Systems* 146: 111–119.

Feintrenie, L., Chong, W.K. & Levang, P. 2010. Why do farmers prefer oil palm? Lessons learnt from Bungo district, Indonesia. *Small-scale forestry* 9(3): 379–396.

Gnych, S.M., Limberg, G. & Paoli, G. 2015. *Risky business: Motivating uptake and implementation of sustainability standards in the Indonesian palm oil sector*. Bogor: CIFOR.

Fairhurst, T.H. & Griffiths, W. 2014. *Oil Palm: Best Management Practices for Yield Intensification*. Singapore: International Plant Nutrition Institute (IPNI).

Fauzi, A. & Oxtavianus, A. 2014. Pengukuran Pembangunan Berkelanjutan di Indonesia. *Jurnal Ekonomi Pembangunan* 15(1): 68–83.

Haryono, Sarwani M., Ritung S., et al. 2011. *Peatland Map of Indonesia*. Bogor: Center for Research and Development of Agricultural Land Resources, Agricultural Research and Development Agency, Indonesia Ministry of Agriculture.

Hettig, E., Lay, J., & Sipangule, K. 2016. Drivers of households' land-use decisions: A critical review of micro-level studies in tropical regions. *Land* 5(4): 1–32.

Hussen, Ahmed M. 2004. *Principles of Environmental Economics second edition*. Taylor & Francis e-Library.

Inovasi Bumi (INOBU). 2016. *A Profile of Oil Palm Smallholders and Their Challenges of Farming Independently*. Jakarta: Inovasi Bumi.
Jelsma, I., Schoneveld, G.C., Zoomers, A. & Van Westen, A.C.M. 2017. Unpacking Indonesia's independent oil palm smallholders: an actor-disaggregated approach to identifying environmental and social performance challenges. *Land Use Policy* 69: 281–297.
Jopke, P. & Schoneveld, G.C. 2018. *Corporate commitments to zero deforestation: An evaluation of externality problems and implementation gaps* (Vol. 181). Bogor: CIFOR.
Koh, L.P. & Wilcove, D.S. 2008. Is oil palm agriculture really destroying tropical biodiversity? *Conservation letters* 1(2): 60–64.
Krishna, V.V., Kubitza, C., Pascual, U. & Qaim, M. 2017. Land markets, property rights, and deforestation: insights from Indonesia. *World Development* 99: 335–349.
Lee, J.S.H., Ghazoul, J., Obidzinski, K., & Koh, L.P., 2013. Oil palm smallholder yields and incomes constrained by harvesting practices and type of smallholder management in Indonesia. *Agronomy for Sustainable Development* 34: 501–513.
Lee, J.S.H., Abood, S., Ghazoul, J., Barus, B., Obidzinski, K. & Koh, L.P. 2014. Environmental impacts of large-scale oil palm enterprises exceed that of smallholdings in Indonesia. *Conversation letters* 7: 25–33.
Lim, C.I. & Biswas, W. 2015. An evaluation of holistic sustainability assessment framework for palm oil production in Malaysia. *Sustainability* 7(12): 16561–16587.
McCarthy, J.F. Zen, Z. 2016. Agribusiness, agrarian change, and the fate of oil palm smallholders in Jambi. In: McCarthy, J.F. Cramb, R. (Eds.). *The Oil Palm Complex: Smallholders and the State in Indonesia and Malaysia*. Singapore: National University of Singapore.
Meijaard, E et al. 2017. *An impact analysis of RSPO certification on Borneo forest cover and orangutan populations*. Bandar Seri Begawan, Brunei Darussalam: Pongo Alliance.
Molenaar, J.W., Orth, M., Lord, S., Taylor, C. & Harms, J. 2013. *Diagnostic study on Indonesian oil palm smallholders. Developing a better understanding of their performance and potential*. Jakarta: IFC.
Pacheco, P., Schoneveld, G., Dermawan, A., Komarudin, H., & Djama, M. 2018. Governing sustainable palm oil supply: Disconnects, complementarities, and antagonisms between state regulations and private standards. *Regulation & Governance*: 1–31.
Peraturan Menteri Pertanian Republik Indonesia Nomor 11/Permentan/Ot.140/3/2015. *Sistem Sertifikasi Kelapa Sawit Berkelanjutan Indonesia* (Indonesian sustainable palm oil certification system).
Potter, L. 2016. How can the people's sovereignty be achieved in the oil palm sector? Is the plantation model shifting in favour of smallholders? In J.F. McCarthy & K. Robinson (eds), *Land and development in Indonesia: searching for the people's sovereignty*: 315–342. Singapore: ISEAS-Yusof Ishak Institute.
Prokopy, L.S., Floress, K., Klotthor-Weinkauf, D. & Baumgart-Getz, A. 2008. Determinants of agricultural best management practice adoption: Evidence from the literature. *Journal of Soil and Water Conservation* 63(5): 300–311.
Rietberg, P.I. & Slingerland, M. 2016. *Barriers to smallholder RSPO certification. A science-for policy paper for the RSPO*. Wageningen: Wageningen University.
RSPO. 2012. *Buku panduan penerapan prinsip dan indicator RSPO untuk petani kelapa sawit*. Jakarta: RSPO.
Sahara, H & Kusumowardhani, N. 2017. *Smallholder finance in the oil palm sector: Analyzing the gaps between existing credit schemes and smallholder realities* (Vol. 174). Bogor: CIFOR.
Saragih, B. 2017. *Oil Palm Smallholders in Indonesia: Origin, Development Strategy and Contribution to the National Economy*, presented at the World Plantation Conference 18–20 October 2017, Jakarta.
Sumarga, E., Hein, L., Hooijer, A. & Vernimmen, R. 2016. Hydrological and economic effects of oil palm cultivation in Indonesian peatlands. *Ecology and Society* 21(2): 52.
Surahman, A. Soni, P. & Shivakoti, G.P. 2018. Are peatland farming systems sustainable? Case study on assessing existing farming systems in the peatland of Central Kalimantan, Indonesia. *Journal of Integrative Environmental Sciences* 15(1): 1–19.
Tietenberg, T. & Lynne, L. 2009. *Environmental and Natural Resources Economics* 8 editions. New York: Pearson.
Wandemberg, J.C. 2015. *Sustainable by Design*. South Caroline: Create Space Independent Publishing Platform.
Winrock & SPKS. 2018. *The key characteristics of independent smallholders in the context of sustainable palm oil*. Jakarta, Indonesia: Winrock.

Rural Socio-Economic Transformation – Kinseng et al. (Eds)
© 2019 Taylor & Francis Group, London, ISBN 978-0-367-23603-8

Farmer households' vulnerability and coping strategies of floods in Kertamulya Village, Karawang District, Indonesia

S. Brigita & M. Sihaloho
Department of Communication and Community Development Sciences, Bogor Agricultural University, Bogor, West Java, Indonesia

ABSTRACT: Natural and environmental disasters such as flooding have direct impacts on farmers' lives. This condition drives farmers to have strategies in coping with flood while maintaining their livelihoods. The objective of this paper is to see the correlation between the level of livelihood vulnerability and the level of resilience of farmers exposed to floods with the livelihood assets owned. Using purposive sampling, 45 respondents was selected from Kertamulya Village, Karawang District to answer questionnaires. The results showed that the livelihood asset affects the level of vulnerability and resilience. The lower the livelihood asset, the more vulnerable the households in facing floods.

1 INTRODUCTION

A large number of Indonesian people who work in the agricultural sector makes Indonesia dubbed as an Agricultural Country. Amanah (2014) defines farmer households as part of the farmer community and people who works in non-farm sector (double income pattern). Farmer households have the following characteristics, namely humans who live together, interact, and cooperate for a long time. The farmer household is an institution or organization in the form of a family that fulfill their daily needs by farming.

In implementation, there are many factors that influence the lives of farmers, one of which is the factor of natural and environmental sustainability. Floods are a manifestation of natural disasters that can have an impact on the lives of farmers. The concept of flooding according to Kodatie & Sugiyanto (2002) is that the excess water that inundates an area that is normally dry occurs as a result of the capacity of the river not being able to accommodate water flowing on it or the excess of local rainwater. The excess of local rainwater that causes flooding can be caused by two things, namely the saturation of the land in that place and the still high water level in the river channel. High soil saturation will cause the level of soil absorption (infiltration) to be low so that surface runoff becomes high.

A large area of land in Kertamulya Village is used for rice fields of 313,000 Ha so that many residents have a livelihood as farmers. Facing a flood situation makes the farmers household make a living strategy to survive the situation. The floods in Kertamulya village have occurred at least the last 10 years, according to what was obtained from in-depth interviews with local stakeholders, community leaders and the community. Floods that occur generally attack community-owned agricultural land. Floods disaster annual basis in the rainy season or the second planting season. Floods destroyed the farmer's household rice fields. This is important to discuss how local people are looking for livelihood strategies to deal with that crisis situations. Because most of the Kertamulya Village area is rice fields and makes the most chosen livelihoods as a farmer. Kertamulya Village, Pedes Subdistrict, Karawang District is is flowed by several streams such as the Citarum River and the Cisoga River. The Kertamulya village is the meeting point of the two rivers, so that when the rainy season comes it can cause the floods that will have an impact on agricultural land.. Especially a few years before the dredging times made the flood more severe. As a result, many rice fields were damaged, farmers failed to harvest.

Dharmawan (2007) said that livelihood strategies are tactics and actions created by individuals or groups in order to maintain life or livelihood strategy. The choice of livelihood strategy is largely determined by the availability of resources and the ability to access these livelihood resources. According to Ellis (1999), there are five types of basic capital in livelihood assets that can be owned or controlled by households to achieve their livelihood, namely natural capital, human capital, physical capital, social capital, and financial capital.

Scoones (1998) divides the classification of livelihood strategies that may be carried out by farmer households, namely: (1) engineering livelihoods of agriculture, which is carried out by utilizing the agricultural sector effectively and efficiently both through the addition of external inputs such as technology and labor work (intensification), or by expanding arable land (extensification); (2) multiple livelihood patterns, which are carried out by applying diversity of livelihoods by finding work other than agriculture to increase income, or by mobilizing family labor (father, mother and child) to work in agriculture and earn income (diversification of livelihood); and (3) spatial engineering is an effort carried out by means of permanent/circular mobilization/migration of population (migration) in order to find livelihood sources elsewhere.

Various efforts to make livelihood strategies are carried out by farmers' households to survive against the vulnerability. The condition of farmland owned by farmers in Kertamula Village exposed to flooding during the rainy season makes a number of farm households have a high level of resilience in facing vulnerability. Adger (2000) said that resilience is the ability of groups to overcome external pressure as a result of social, political and environmental changes. The concept of resilience is a broad concept, which includes capacity and ability to respond in crisis or emergency situations. A crisis condition can be experienced by farm households due to crop failure caused by unpredictable natural conditions.

Sunarti (2013) said that resilience is the ability to adapt to the change, especially unwanted changes that bring problems and even cause crisis conditions. As a farmer's households in Kertamulya Village must adapt when the floods come. Resilience as a form of adaptation or defense from a person or society in the face of a condition. The level of resilience can be measured by various indicators. One of them is measuring the level of resilience based on the time needed by the household for recovery when a crisis situation occurs, such as the results of Azzahra (2015) study.

The farmer's household did various efforts to fulfilling life necessities also survived when vulnerability. The vulnerability is a common concept in climate change research as well as in research communities related to natural disasters and disaster management, ecology, public health, poverty and development, safe livelihoods and hunger, sustainable science, and land change (Fussel 2007).

Based on the above explanation, there is a connection between livelihood strategies, livelihood structure, capital income, resilience, and learning. Hence, it is very important and interesting to study how the livelihood strategy of farmers household in Kertamulya Village to survive the vulnerable condition that occurs due to flooding.

There are four main objectives of this research: (1) to analyze the livelihood structure and the livelihood strategy of farmer households in exposed floods areas; (2) to analyze the influence of Livelihood Assets to the Level of Livelihood Vulnerability; (3) to analyze the influence of Livelihood Assets to the Level of Resilience; and (4) to analyze comparison of the influence between Level of Livelihood Assets and Level of Vulnerability to the Level of Resilience of farmer households in Kertamulya Village.

2 METHODS

This research was conducted in Kertamulya Village, Pedes Sub-District, Karawang District, West Java Province in August 2017-January 2018. Using quantitative research methods supported by qualitative data to enrich data and information obtained in order to understand social phenomena that occur in the field. This quantitative approach is supported by data collection methods in the form of numbers obtained from measuring instruments in the form of questionnaires. Qualitative data is needed to retrieve descriptive data in the form of social symptoms obtained through in-depth interviews with informants.

Determination of respondents with taking a sample of 45 respondents. The selection of respondents used purposive sampling technique. Selected along with the head of the union of farm laborers (Ketua Gapoktan) and heads of farmer groups who know exactly who the farmer households are in the flood fields. This happened because not all farmers in Kertamulya village were flooded. The informants in this study were chosen intentionally (purposive) and the selected numbers were not locked. Determination of this informant will be done using the snowball technique that allows the acquisition of data from one informant to another informant.

The quantitative data were collected using a structured questionnaire. While qualitative methods were collected by using in-depth interviews. This method known in various terms and has been used, such as (Kinseng et al. 2014) or field research (Harahap & Dharmawan, 2018). In-depth interviews were carried out by around 40 people including local stakeholders such as the village head, village secretary, chairman of the farmer organizations, UPTD Local government. The elite figure, religious figure and the community of Kertamulya Village were involved in extracting information from in-depth interviews. Even people from the Neighborhood Village (Malangsari Village) also helped collect data taken through in-depth interviews. Quantitative data obtained from questionnaires were then processed by using the Microsoft Excell 2010 application and statistics for social sciences (SPSS) 22.0 for Windows and Minitab 16.0 applications. The statistical test that will be used is Linear Regression to determine the effect of the variables to be studied. Stepwise regression tests are used to see differences between variables. Data collection techniques use interview techniques and qualitative data analysis, namely data reduction, data presentation, and verification to support quantitative and quantitative data. Adger (2006) said that the general level of livelihood known as LVI (Livelihood Vulnerability Index) can measure exposure, sensitivity, and adaptive capacity. This study uses the LVI calculation formula which based on the IPCC (Intergovernmental Panel of Climate Change) is LVI - IPCCd = (Ed - Ad) × Sd. Ed is exposure, Ad is an adaptive capacity, and Sd is sensitiveness.

3 RESULTS AND DISCUSSIONS

3.1 *General description of Kertamulya Village*

Kertamulya Village, Pedes Subdistrict, Karawang District, West Java Province is located in the northern coastal area of Karawang District which administratively has an area of approximately 529,170 hectares. Kertamulya Village consists of four hamlets and has 8 RW (Head of Hamlets) and 16 RT (Neighbourhood). Each hamlet consists of one RW and two RT. The four hamlets are as follows: (1) Suka Jaya Hamlet consists of 2 RW and 4 RT, has 1816 residents; (2) Sukarela Hamlet consists of 2 RW and 4 RT, has 1455 residents; (3) Jayamukti Hamlet consists of 2 RW and 4 RT, has 1244 residents; (4) Cintasari Hamlet consists of 2 RW and 4 Rs, having 1969 residents. The population of Kertamulya Village based on village monograph data (2016) is 6471 people. Consisting of 2254 family heads.

Most of the residents of Kertamulya Village are indigenous people, and some of them are migrants. The total population of Kertamulya Village is 6471 people consisting of 3202 men and 3269 women. The value of mutual cooperation is still highly respected by the community, The society of Kertamulya Village like to help each other. Kertamulya village has two irrigation sources for rice fields, namely Cisoga and Citarum rivers. Both of them can overflow, especially during the rainy season, so that it can cause flooding at some points in the area of rice fields owned by residents.

3.2 *The livelihood structure of farmers' households of Kertamulya Village*

There are 45 households who were respondents in this research, then classified into three strata based on the classification of farmers' income according to BPS in 2008. The three class are the lower class, the middle class, and the upper class. The total income in the question is a net income, a combination of income from the on-farm sector, the off-farm sector, and the non-farm sector. The different household income of farmers makes a variation in the level class of farmer household income. The on-farm structure dominates in all class. The lower class tends to have very low income. The lower class only relies on non-farm structures even the value tends to be negative at IDR 6.8 million (deficit). Income of farmer household in three class more higher by on farm sector because even though the area of land owned by the household is not as much as the upper class, there so many middle class of farmer households who make efforts to lease or pledge agricultural land in order to increase the income.

The structure of non-farm income is also quite high in the middle and upper class. Existing floods make the farmer households unable to utilize agricultural products optimally. So that many households have a strategy to work outside the agricultural sector and increase their income in the non-farm sector. The most structure off-farm is utilized by the lower class of farmer households. There so many farmers household in the lower class use the off-farm sector by working as farm laborers on other people's land.

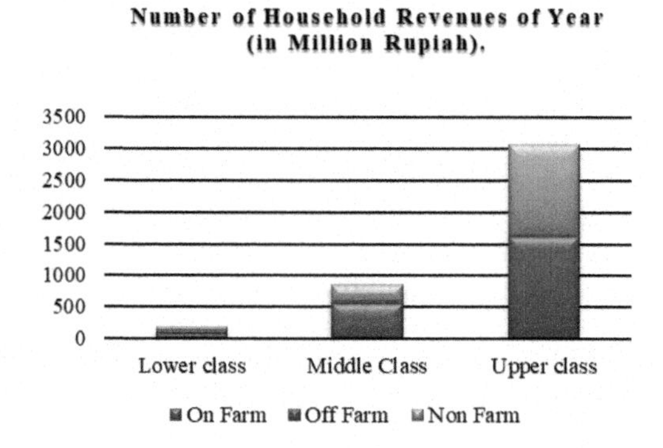

Figure 1. The amount of farmer household income based on the economic class of Kertamulya Village in 2016–2017.

Farmer's households of the middle class rely heavily on farm structure and the non-farm structure. They make a strategy to share the results of rice fields, pawn and also rent land. Non-farm income obtained based on employment outside the agricultural sector such as opening a business selling food, opening stalls and other livelihood activities outside the agricultural sector. Farmers households in the upper class rely on farm structure because they have a high capital, namely the amount of extensive agricultural land. There so many non-farm structures are also carried out because they have a strong financial. For example, the financial capital is used to open stores, and other jobs.

Based on Figure 2, the saving capacity in every class of farmer's households varies. The upper class has the highest income because most farm households belonging to the upper class have a large amount of land, besides that many farm households in the upper layers have other sources of income and business besides farming. The level of household expenditure of farmers in the lower class is more alarming because it tends to be a deficit. Seeing the level of expenditure in every class of farmer households shows that they are difficult to have the saving capacity. The farmer's household of Kertamulya Village as a whole stated that they did not have a savings amount or saving capacity. They assume all the income earned will surely be released immediately for needs.

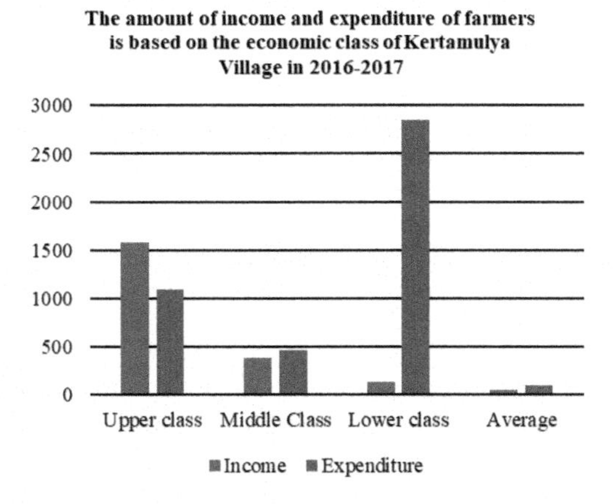

Figure 2. The amount of income and expenditure of farmers based on the economic class of Kertamulya Village in 2016–2017.

3.3 *The livelidhood strategy of farmers' households of Kertamulya Village*

Floods have a significant impact on some farmers in Kertamulya Village. According to the local community, Kertamulya Village has two types of paddy fields namely flood-free rice fields and cow fields. Sawah kobak is the term of the local community about rice fields where the land is very low, like a basin, so that if paddy water flows from upstream, it will be difficult to flow back downstream so that the water soaks a number of low-lying rice fields. Various strategies livelihoods are made to survive.

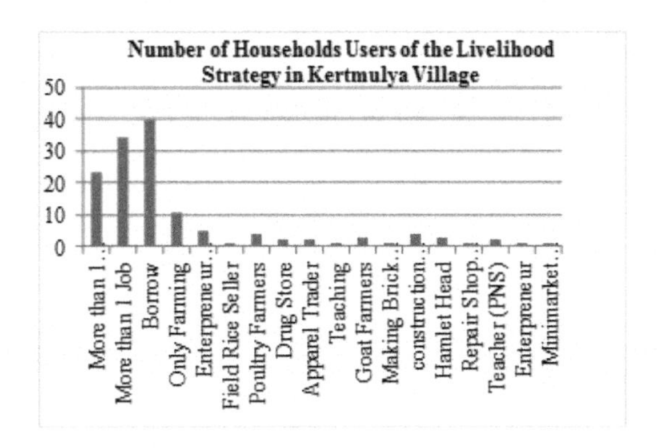

Figure 3. Number of households users of the livelihood strategy in Kertmulya Village.

Most farmer households diversify their livelihoods. Diversification of livelihood by mobilizing human capital owned or involving household members such as wives, and children to be engaged in the livelihood. In addition, livelihood diversification is also carried out by conducting a series of livelihood activities as additional work to increase livelihoods. Doing work with an off-farm structure is a choice that is quite chosen by households. This is because the household has other skills and expertise outside the agricultural sector which can be done in an effort to earn income. Besides that, the borrowing livelihood strategy is also the most preferred. The lower and middle layers tend to make strategies to borrow from relatives or neighbors by relying on social capital or a sense of family owned. The upper layer tends to borrow from conventional banks because the nominal borrowed tends to be higher because it is intended for farming capital.

3.4 *Analysis of livelihood assets of farmers' households in Kertamulya Village*

Farmer's households in Kertamulya Village have diverse and different livelihood capital in each economy class. Stratification based on income level results in three economic classes, namely the lower class, the middle class, and the upper class. The flood disaster in Kertamulya Village was an annual flood that occurred once, namely in the second planting season.

Livelihood capital consists of five capital namely physical capital, financial capital, human capital, natural capital, and social capital. Physical capital referred to in the study of household assets both assets on the farm and non-farm assets owned by households. Financial capital consists of savings and loans.

Human capital in this study includes the amount of labor allocation, education level, and skill level possessed by the head of the family. Natural capital in this study covers the area of land used by farmers to support livelihood activities and the level of farmer household access to a number of lands. Social capital in research is seen from the level of organizational participation.

The livelihood activities of farmers household are influenced by the utilization of livelihood capital owned. This happens because it depends on the situation or situation faced by the farmer's household. If there is a crisis situation, it will give a disruption to the livelihood capital so that farm households have the potential to experience vulnerability.

Figure 4 shows a comparison of the utilization of livelihood capital used by each class of farmer households. Households in the lower class have an average gross income of around Rp14.4 million. Farmershousehold in the lower class tends to use social capital and financial capital because the two capital values are quite high. Social capital is high because most of the farmers' households in the lower levels follow community organizations in the village such as farmer groups, recitation groups, and others. Social capital is used with the existence of relationships and kinship based on household involvement in organizations in the village. The organizations that are most followed are the study groups. Although the recitation group did not provide direct assistance during a flood, the kinship among its members made a high mutual help and helped in lending. Almost all farmer households in Kertamulya Village are not actively participating in existing farmer groups. Farmer households are indeed registered in farmer organizations in the village, but the notion of the existence of farmer groups is not felt because it has long been a vacuum. The lack of clarity in the existing organizational structure makes the existing farmer organization tend to be passive compared to farmer groups in other villages. High financial capital is seen based on lend indicators, which of course are carried out by farmer's households to survive. Farmer household in lower class have limitations in the utilization of human capital, physical capital, and natural capital.

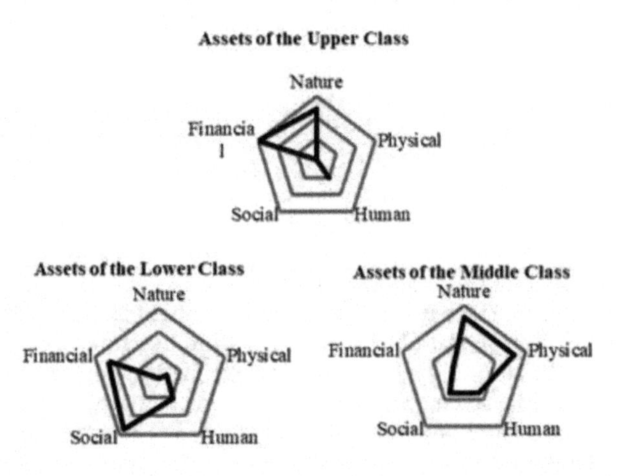

Figure 4. Pentagonal image of utilization of livelihood assets between the farmers household class in the village of Kertamulya in 2016–2017.

The middle class has an average net income of 37.15 million rupiahs. The use of capital in the middle class tends to be high on natural capital and physical capital. Natural capital is quite high because most of the farmer's households in the middle class have agricultural land as an indicator of natural capital, ranging from private fields, pawn, or rent. Physical capital owned by farmer households in the middle layer is quite high with the ownership of on-farm physical capital and non-farm physical capital.

The upper class is farmer households that have an average net income of 129.4 million rupiahs, is tended to be dominant in utilizing financial capital and natural capital, both of which are in the high category. Of course, because many farmer households in the upper class have private rice fields and financial conditions that tend to be stable.

3.5 *Livelihood Vulnerability Index (LVI) in Kertamulya Village*

Livelihood Vulnerability Index of farmer households in Kertamulya Village is 0.15 which indicates that farmer households in Kertamulya Village are categorized as vulnerable due to floods that occur regularly every year. The value of LVI is based on exposure (0.64), sensitivity (0.74), and adaptive capacity (0.44). Based on three indicators of vulnerability, namely exposure, sensitivity, and adaptive capacity when viewed based on the results of calculations, the value of adaptive capacity is the lowest among the other indicator values.

This occurs because the level of adaptive capacity owned by low-income households sees one factor in the level of adaptive capacity, namely the frequency of the gathering farmer groups and ownership of reserve savings, as well as the existence of farm capital loans. Many farmer households stated that farmer organizations in the village almost never held activities or meetings. The existence of farmer groups in the local area is felt to be less active by most farm households. This makes many farm households say that it is very rare to hold farmer group meetings or associations at certain times. In addition, the low ownership of reserve savings and low capital loans make the value of the adaptive capacity of farmer households to be of low value.

While the exposure value of 0.64 indicates that farmer households in Kertamulya Village are very exposed. Exposure factors are seen based on the distance of the river atayu irrigation with agricultural land, the percentage of paddy fields affected by floods, and the percentage of paddy fields damaged by floods. Some farm households stated that many of their paddy fields were damaged due to the flood disaster. This results in reduced yields and quality.

Table 1. Livelihood Vulnerability Index in Kertamulya Village on 2016–2017.

Variable	Sub Component	Sub Component Value	Main Component Value
Exposure	the distance between land and river	0.57	0.64
	percentage of land affected by flooding	0.70	
	percentage of land damaged by flooding	0.67	
Sensitivity	Alternative land ownership	0.74	0.74
	reserve savings	0.74	
	pesticide subsidies	0.79	
	fertilizer subsidies	0.70	
Adaptive Capasity	irrigation repair ability	0.74	0.44
	increase in post-flood land fertility	0.71	
	the frequency of the time organizations gather	0.39	
	the frequency of organizations gathered each month	0.26	
	the frequency of counseling every year	0.25	
	ownership of reserve savings	0.37	
	the existence of a capital loan	0.37	
	Livelihood Vulnerability Index		0.15

3.6 *Factors affecting the LVI in Kertamulya Village*

Based on the description above shows that natural capital and physical capital influence the level of vulnerability of household livelihoods of farmers who experience flooding in Kertamulya Village. So that the regression equation is obtained:

$$LV = 1.502 - 0.400C_n + 0.693C_p - 0.003C_h + 0.036C_s + 0.063C_f \quad (1)$$

where LV = level of vulnerability; C_n = nature capital; C_p = physical capital; C_h= human capital; C_s= social capital; and C_f = financial capital.

Table 2 shows that there is two influential capital in determining the value of LVI, namely natural capital and physical capital. Natural capital is measured based on capital variables in the form of land used by farmer households to support their livelihood activities, the ability of farmers to carry out agricultural and gardening activities, the ability of farmers to manage their own land, and the ability of farmers to work and manage fields or land other people.

Farmer's households with low capital ownership will make their vulnerability high. Because the amount of land is limited, or the inability of farmers to own their own land will produce fewer crops and generate lower income. Conversely, for households that have high livelihood capital, it will be very helpful when facing a crisis situation. Household farmers who have high natural capital will make their vulnerability lower. The existence of a number of alternative land or land owned outside the flood area also helps households to continue to meet their needs even in conditions of crisis.

Physical capital measured based on capital ownership variables in physical form, both from agricultural capital and non-agricultural capital. The physical capital of the farm consists of hoes, pest spraying machines, tractor machines, rice grinding machines. Whereas non-agricultural physical capital consists of televisions, refrigerators, washing machines and motorbikes. Physical capital can be utilized by making a strategy for new livelihoods, selling or mortgaging some of the physical capital they have. The hope is that by relying on this strategy, households will still be able to survive in a crisis situation and still be able to meet their household needs and needs. Like household farmers who use non-agricultural physical capital such as refrigerators used to make ice cubes and sell them, motorbikes are used to save on agricultural capital costs and can be used to work as a driver of motorcycle taxis (ojek). Physical capital of agriculture such as hoes, pest spraying machines, tractor machines owned by farmers is very helpful in reducing the cost of agricultural capital. This really helps make farmer households spend as little capital as possible but still get maximum results. The low level of households that have non-agricultural capital will affect the level of vulnerability. Households that have high physical capital will make household vulnerabilities relatively low.

3.7 *The resilience of farmers' households in Kertamulya Village*

The time of farmer's household recovery when a crisis occurs is used as an indicator to measure the level of resilience. The linear regression analysis test was used to test the factors that

Table 2. Factors affecting the LVI of farmers' household in Kertamulya Village 2016–2017.

	Unstandardized Coefficients		Standardized Coefficients		
Model	B	Sted.error	Beta	t	Sig
(Constant)	1.502	.562		2.672	.011
Nature Capital	**-.400**	**.196**	**-.288**	**-2.045**	**.048**
Physical Capital	**.693**	**.299**	**.359**	**2.319**	**.026**
Human Capital	-.003	.030	-.014	-.097	.923
Social Capital	.036	.067	.085	.540	.592
Financial Capital	.063	.159	.070	.396	.694

a. *Dependent Variable*: Level of Vulnerability (LVI)

Table 3. Factors affecting the resilience of farmer households in Kertamulya Village on 2016–2017

	Unstandardized Coefficients		Standardized Coefficients		
Model	B	Sted.error	Beta	t	Sig
(Constant)	2,964	1.104		2.684	.011
Nature Capital	-.679	.384	-.269	-1.766	.085
Physical Capital	.331	.587	.095	.565	.575
Human Capital	.044	.059	.115	.746	.460
Social Capital	.260	.131	.340	1.988	.054
Financil Cpital	-.652	.311	-.403	-2.094	.043

a. *Dependent Variable*: Level of Resilience

influence the level of resilience. Alpha is used at 0.2 or 20 percent, which means that the tolerance of errors in the regression test is 20 percent, and the true value is 80 percent.

Natural capital which consists of extensive land ownership used by farmer households in their livelihood activities and the level of access to natural resources (land) has an influence on the level of resilience. Many or considerations of the time needed by farmers to survive in times of crisis are also seen based on natural capital owned. The more ownership of natural capital or the existence of alternative land owned by farmer households, or the greater the level of access of farmer households to livelihood activities will make agricultural households immediately rise from the crisis. Farmers households provide alternative rice fields in addition to the rice fields provided in Kertamulya Village, which makes them have others. Alternative fields are needed to make the level of resilience high, which means that the farmer household does not need to linger on the problem of the crisis because he immediately arises from the problem of crisis. The longer the recovery time the farmers have, the lower the level of resilience they have. At present farmers only rely on one land needed during a flood, this will divert farm households to increase their household income and make farmers need a lot of time to survive. Because the rice fields owned in Kertamulya Village are cow fields (prone to flooding and including tubs). If farmer households only focus on one rice field, they will not have another source, if the Kobak rice fields are damaged by floods, farmers will decline.

Social capital is also one of the livelihood capital that influences the level of resilience of household farmers in Kertamulya Village. The high level of kinship in Kertamulya Village greatly helps the community to survive in various situations that occur. Social capital is one of the capitals which has an interest in the level of resilience. Social capital is seen based on household participation in a village organization. The more houses that accept the organization, the relationship will be higher. The high level of kinship in Kertamulya Village is also one of the strategic social capital when there is a crisis. The more households that participate in an organization, the higher their relationships and kinship values than households that need to be inactive in any organization and to increase collectively in their environment. The high sense of kinship created in the Kertamulya Village community was very helpful when the flood incident came. The higher the social capital that is needed, it will help farm households in carrying out activities to provide mutual assistance and help each other when the floods come. This will make farm households not linger in times of crisis. Recovery of the time needed by farmer households is small and shows a high level of resilience.

The last capital that affects the level of resilience is financial capital. Financial capital has a sig value of 0.043. Financial capital is measured based on the level of savings held by the farmer's household and the amount of income and loan strategies that can be made by farmer households. Based on the linear regression test in Table 3 shows that financial capital has an influence on the level of resilience. the higher the financial capital used by farmer households, the higher the level of resilience. The more financial capital owned by farmer households will make the recovery time of farmer households less. Households that have high livelihood capital will make the household do not need a long time to return business loan capital or lend money to meet household needs. Financial capital owned by farmer households will facilitate recovery because it is able to repay loan funds quickly, easily obtain loan funds, assist in making entrepreneurship. The high financial capital that is owned also makes farmer households do not need a long time to find replacement jobs when they lose their source of income because the financial capital they have can be used as their own business such as opening stalls, opening drug store and the type of business they own. So that the regression equation is obtained:

$$LR = 2.964 - 0.679C_n + 0.331C_p + 0.044C_h + 0.260C_s - 0.652C_f \tag{2}$$

where LR = level of resilience; C_n = nature capital; C_p = physical capital; C_h= human capital; C_s= social capital; and C_f = financial capital.

3.8 *The effect between the level of livelihood assets and level of livelihood vulnerability towards the level of resilience of farmers' households*

The influence test was carried out using stepwise regression test. The stepwise regression test was carried out to see the comparison of the effect of the level of living assets and the level of vulnerability of farmer household livelihoods in the Kertamulya village which was flooded to the level of resilience of the farmer's household income. This regression test uses the alpha of 20 percent or value 0.2 which means that the tolerance of errors in the regression test is 20 percent and the truth is worth 80 percent.

Based on the results of the stepwise regression test it can be seen that the level of livelihood assets is more influential on the level of livelihood resilience compared to the level of livelihood vulnerability. The level of asset livelihood is measured based on the level of ownership of natural capital, physical capital, human capital, financial capital, and social capital. The level of livelihood assets is more dominant affecting the level of resilience. Based on the results of stepwise regression, it shows three capital that has a significant influence on the level of resilience, namely social capital, financial capital, and natural capital. The level of vulnerability is measured based on exposure level, adaptive capacity level, and sensitivity level. Based on the results of the stepwise regression test, the level of livelihood vulnerability also affects the level of livelihood resilience. But only the sensitivity level has a significant influence. Sensitivity levels are measured based on the existence of reserve savings, ownership of alternative land, and availability of subsidies for fertilizers and pesticides for farm households.

Table 4. The effect of livelihood assets and level of livelihood vulnerability towards the level of resilience of farmers' households

Step	1	2	3	4
Constant	62.87	90.16	-15.21	-204.53
Financial Capital	85	132	144	140
T-Value	2.12	2.98	3.28	3.30
P-Value	0.040	0.005	0.002	0.002
Social Capital		-45	-42	-48
T-Value		-2.16	-2.02	-2.37
P-Value		0.037	0.049	0.023
Level of Sensitivity			39	43
T-Value			1.76	1.99
P-Value			0.086	0.053
Nature Capital				122
T-Value				1.95
P-Value				0.058
S	86.1	82.7	80.7	78.1
R-Sq	9.47	18.89	24.22	30.81
R-Sq (adj)	7.36	14.61	18.67	23.89
Mallows CP	9.6	6.6	5.4	3.7

Based on the results of the stepwise regression test the equation is as follows:

$$LR = -205 + 140C_f - 47.7C_s + 122C_n + 43.0LS \tag{3}$$

where LR = level of resilience; C_n = nature capital; C_p = physical capital; C_s= social capital; and LS = level of sensitivity

4 CONCLUSIONS

Flooding had a large impact on farmer's households in Kertamulya Village. Regular annual floods make the farmer's households need to be able to withstand vulnerabilities. Based on the Livelihood Vulnerability Index, Kertamulya Village was classified as vulnerable due to the flood disaster, because flooding attacked the rice fields which became a source of income for most of the Kertamulya Village people who work as farmers. Farmer households must be alert to the flood situation, especially the floods that have occurred even more than the last 10 years to date. Various livelihood strategies are carried out in order to survive. Working with relying on farm structure is mostly done considering that most people work as farmers. Working in the non-farm sector is also an alternative to supporting farmers' household income to survive. By relying on various expertise and capital to run other businesses outside of agriculture. Many farmer's households have other jobs besides working as farmers.

Livelihood diversification is also done by relying on household members to work. Many farmer's households have children or wives who also work to increase their income. Livelihood assets are also used as a way to increase income and help deal with floods. The strategy of natural capital by having a very influential alternative reserve area helps farmers face a flood situation. Most of their rice fields in the village of Kertamulya experienced flooding, thus encouraging them to have rice fields outside their village. Physical capital is also used as another strategy by maximizing the number of items such as refrigerators to be used as an effort to sell ice cubes, motorbikes for the business of motorcycle taxi drivers, complete physical capital to reduce farm capital. Likewise with financial capital is the capital most felt by households helping in the face of flooding. With the availability of financial capital to help them feel safe in returning loan money, get easy access to capital loans, and buy household needs. Talking about flooding makes farm households have a high degree of resilience in order to be able to withstand floods. The amount of exposure to rice fields due to flooding, the level of sensitivity of households in the face of flooding, as well as the adaptability of household farmers affect the survival of farmer households. Similar to the increasing number of assets that are owned will encourage farmers to quickly recover from flood conditions.

But in fact, there are still a lot of farmer households that are actually minimal in the ownership of livelihood assets. Especially the ownership of rice fields, most farmers in Kertamulya Village are small farmers who do not own the fields. They work on other people's fields. Whereas farmer households in the dominant upper layer own agricultural land. From the aspect of social capital many households actively participate in community organizations, but unfortunately, these organizations tend to be passive and not running. As existing organization pharmacies tend to be passive, many farmers do not feel the benefits because of the lack of resources that are willing to organize and organize farmers. So that many farm households feel unhelpful from the side of the organization. But they are strong at optimizing kinship values and existing family values. They help each other and provide assistance when the flood comes by cleaning up the shared irrigation channels or giving loans to one another. Seeing the routine flooding that occurs also makes farmer households take the initiative to shift their planting time so that their rice yields are better than the results of flood-affected rice. Because the results of rice exposed to floods tend to be black, few, and make selling prices drop dramatically.

REFERENCES

Adger, W. N. 2000. Social and ecological resilience: are they related? *Progress in Human Geography* 23 (3): 347–364.
Adger, W. N. 2006. Vulnerability. *Global and Environmental Change* 16:268–281.
Amanah, S. 2014. *Pemberdayaan sosial petani-nelayan, keunikan agroekosistem, dan daya saing*. Jakarta: Yayasan Pustaka Obor Indonesia.
Azzahra, F & Dharmawan, A. 2015. Pengaruh livelihood assets terhadap resiliensi nafkah rumah tangga petani pada saat banjir di Desa Sukabakti Kecamatan Tambelang Kabupaten Bekasi. *Sosiologi Pedesaan Journal* 3(1): 1–9.
Dharmawan, A. 2007. Sistem penghidupan dan Nafkah pedesaan pandangan sosiologi nafkah (livelihood sociology) mazhab barat dan mazhab bogor. *Sodality Jurnal Transdisiplin Sosiologi, Komunikasi dan Ekologi Manusia* 1(2): 169–192.
Ellis, F. 1999. *Rural livelihood diversity in developing countries: Evidence and policy implications*. United Kingdom: Overseas Development Institute Portland House.
Ellis, F. 2000. *Rural livelihood and diversity in developing countries*. London: Oxford University Press.
Fridayanti, N & Dharmawan, A. 2013. Analisis struktur dan strategi nafkah rumahtangga petani sekitar kawasan hutan konservasi di Desa Cipeuteuy, Kabupaten Sukabumi. *Jurnal Sosiologi Pedesaan*, 1(1): 26–36.
Fusel, H. M. 2007. Vulnerability: a generally applicable conceptual framework for climate change research. Global environmental change. *International Journal of Geosciences* 17:155–167.
Harahap, T & Dharmawan, A. 2018. Strategi nafkah dan pemanfaatan relasi-relasi sosial rumah tangga petani kelapa sawit. *Jurnal Sains Komunikasi dan Pengembangan Masyarakat* 2(3): 383–402.
Hartini, Hadi, Sudibyakto, Poniman. 2015. Risiko banjir pada lahan sawah di semarang dan sekitarnya. *Majalah Ilmiah Globe* 17(1): 51–58.
Kodatie, J & Sugiyanto. 2002. *Banjir, beberapa masalah dan metode pengendaliannya dalam perspektif lingkungan*. Yogyakarta: Pustaka Pelajar.
Kinseng, R., Sjaf, S & Sihaloho, M. 2014. Class, income, and class consciousness of labor fishers. *Journal of Rural Indonesia* 2(1):94–104.
Scoones, I. 1998. Sustainable rural livelihood a framework for analysis IDS working paper 72. Retrieved from https://www.staff.ncl.ac.uk/david.harvey/AEF806/Sconnes1998.pdf at 11 December 2018
Sunarti, E. 2013. *Potret ketahanan keluarga Indonesia perspektif keragaman pola nafkah keluarga*. Jakarta: CV Widyalika Utama.
Tommi, Barus, B & Dharmawan, A. 2016. Pemetaan kerentanan petani di daerah dengan bahaya banjir tinggi di Kabupaten Karawang. *Majalah Ilmiah Globe* 18 (2):73–82.

Rural Socio-Economic Transformation – Kinseng et al. (Eds)
© 2019 Taylor & Francis Group, London, ISBN 978-0-367-23603-8

Community adaptation on ecological changes through urban farming innovation for family food security

Sumardjo
Department of Communication and Community Development Sciences, Faculty of Human Ecology, Bogor Agricultural University, Bogor, West Java, Indonesia

A. Firmansyah
Center for Alternative Dispute Resolution and Community Empowerment, Institute for Research and Community Service, Bogor Agricultural University, Bogor, West Java, Indonesia

Manikharda
SEAFAST Center, Institute for Research and Community Service, Bogor Agricultural University, Bogor, West Java, Indonesia

ABSTRACT: Threats to food security worsened by the increasing of agricultural land conversion into non-agricultural purposes. Thousands of hectares of paddy fields and settlements in Majalengka Regency have changed due to the construction of megaprojects such as construction of airports and hundreds of factories established in nine sub-districts. This research aimed to investigate how the innovations of urban farming could be a solution to the threat of food security and how the community adapts to ecological change in the presence of: (1) land conversion, and (2) innovation of urban farming using participatory action research. The results of this study indicated that there were four types of adaptation to the innovation of urban farming, those were: (1) apathetic, (2) reactive, (3) proactive, and (4) anticipatory type, according to the level of adaptiveness toward the changes. Field assistance facilitated the increase in community adaptation in urban farming.

1 INTRODUCTION

In correspond to the incessant development of infrastructure, it is important to watch out for the increasing threat to food security that associated with the conversion of agricultural land to non-agriculture purposes. Thousands hectares of land for paddy fields and settlements in Majalengka Regency have converted due to the construction of megaprojects such as the Cikampek-Palimanan toll road, the construction of airport that requires around 1,800 hectares of land, and hundreds of manufactures established in nine districts in Majalengka Regency (Hidayat et al. 2017).

Agricultural land conversion is hardly avoidable due to infrastructure construction, which basically occurs because of land usage competition between the agricultural sector and the non-agricultural sector. To control the rate of land conversion, the Government of Indonesia has actually implemented various legal instruments, both laws and government regulations (PP). For example, Law 26/2007 concerning Spatial Planning and Law 41/2007 concerning the Protection of Productive Agricultural Land, PP No. 12 of 2012 and other regulations. The effectiveness of the regulation implementation is interesting subject to study as well as to find the solution of the arising problems (Tindaon 2015).

Yard utilization as a form of urban farming innovation is an alternative resolution in encountering land conversion. Urban farming was defined as an industrial chain that produces, processes and sells food and energy to meet the needs of urban consumers (Bailkey &

Nasr 2000). Urban farming is also carried out as an activity to generate income for farmers, especially for those whose main livelihood is from farming. According to Rahayu et al. (2013), in urban farming, the cultivation process is oriented towards increasing production accompanied by an increase in employment. Urban farming is in line with government programs through the Ministry of Agriculture to improve food security and family nutrition (Putri et al. 2015).

The issue of environmental change as an impact of toll road and airport constructions needs to be addressed by developing alternative solutions to the threat to food security or food crisis in affected communities. In an attempt to overcome the food security threat due to land conversion, yard utilization as a form of urban farming with community empowerment strategy might offer as an alternative solution. Therefore the purpose of this study is to investigate how the innovations of urban farming through yard utilization encounters the threat of food security and how the community adapts to the environmental change in the presence of land conversion and innovation of yard utilization.

2 METHOD

2.1 *Research locations*

The research locations were selected purposively, namely in two villages in Sumberjaya Subdistrict, Majalengka Regency, located in the area directly affected by the construction of the Cipali toll road and indirectly influenced by the construction of the Kertajati Majalengka international airport.

2.2 *Participatory approach*

Community empowerment was established through participatory action research based on three angulation technique. The participatory approach in this study was employed by positioning the land owners as the subjects of developing urban farming innovations, and the field assistants acted as facilitator to change the perception of the real needs to be perceived by the owner of the yard. The field assistants were assigned to stay in the research location among the locals for starting March 2018. They acted as facilitators to empower the community as well as observers and data collectors.

2.3 *Analysis of household food expenditure*

Analysis of household food expenditure patterns was carried out from the average data of nine groups of food items based on the proportion of energy balance according to the energy balance indicator of the expected food pattern (PPH) through questionnaires. The nine groups of food include rice, tubers, animal based foods, plant based foods, vegetables, fruits, cooking oil, legumes and sugar. Whereas, the calculation of the average energy and protein adequacy rate were based on age and gender using the Widyakarya Nasional Pangan dan Gizi (National Food and Nutrition) standard. The comparison between consumption of nutrients and the number of nutritional adequacy recommended was referred to as the nutrition consumption rate based on the Health Center Nutrition Officer's Handbook Ministry of Health (Supariasa et al. 2001). To measure the degree of food security at the household level, cross classification of two indicators of food security was used, namely the share of food expenditure and the adequacy of energy consumption (Maxwell & Frankenberger 1992).

2.4 *Profiling adaptiveness type of the subjects*

The profiling of subject types was performed based on the results of observations and in-depth interviews with the community, both by the field assistants as well as the main researcher. During the community empowerment activities, in the evaluation step the following aspects were diligently observed: education level, the activeness during urban farming related activities, the contribution for the group of urban farming participants, initiative in the community, and performance. From those aspects correlations were made, then we assigned the adaptiveness level of each participant. The correlation was evaluated statistically using Spearman correlation (SPSS, Statistical Program for Social Science, Version 22).

3 RESULTS AND DISCUSSION

Based on our observation, the strategy for developing urban farming in a participatory method turned out to be effective in fostering the needs of the local community and creating a mutual awareness of the market to promote yard products, as well as the existing potency and problems of the resource management, either owned or could be accessed by the community. The strength of this participatory empowerment strategy was in the exploration of potential, problems and expectations that come from the community. This strategy was also achieved by a developing a comparative study to other locations of urban farming that were considered successful. It turned out that the comparative study was a good learning process, particularly in "seeing is believing" as well as obtaining the knowledge directly from the experts as resource person for the development of urban farming. These experiences raised public awareness and innovation of urban farming became a perceived need for change in yard utilization. Further explanation will be described in the following sections.

3.1 *Changes in the community after community empowerment*

Figure 1 shows the physical changes of the local underutilized space and land in employing the urban farming innovation in their yard. By substituting the necessities of plant-based foods, namely vegetables which were originally fulfilled through market purchasing, people can grow the needed plants themselves by urban farming. In addition, for certain

Figure 1. One of the respondent's yard before (left) and after (right) urban farming.

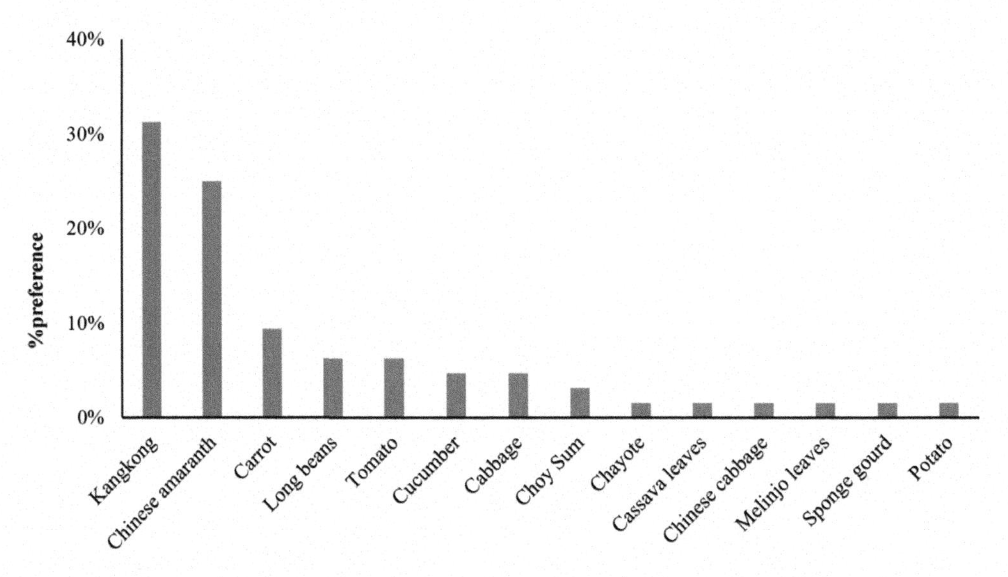

Figure 2. Vegetable preference in the study area.

commodities, the products can be sold, in which the money can be used to fulfill the needs for animal based food.

Based on interviews, observations and FGDs among community members and related figures, an overview of the preferences of the types of vegetables consumed by the community at the research location was obtained. It was observed that the types of plants chosen for cultivation were: kangkong, choy sum, chinese cabbage, tomato, chilis, long beans, beans, and corns. The details can be seen in Figure 2. Determining the right commodity for urban farming innovation was appropriate when the selection was based on the needs of the local community and the possibility to be broadly marketed.

Furthermore, plant maintenance has been carried out by each owner of the yard, including spraying with water enriched with micro-organisms (MOL) that was produced locally, controlling pest, weed and disease, and installation of stakes for plant propagation. In this case, the formation working groups functioned effectively as: (1) a learning media for sharing knowledge and experiences, (2) an economic cooperation body, especially for streamlining input procurement, and (3) as media for synergetic partnership which strengthened the bargaining position in marketing the products in mutually profitable way.

3.2 *Analysis of food security in the household*

After the change occurred in agricultural ecology of the community, food security in the households was considered inadequate. In households with food insecure status, the amount of income became a major factor in achieving food security. As Ariani and Rachman (2003) stated in food insecure household group, the main thing that needed to be done was to improve access to food through increased income followed by enhanced knowledge of food nutrition. Yard utilization contributed in increasing the access to food availability. After application of urban farming, the results of the calculation of the numbers of energy based on food analysis shows the improved status of household and individual nutritional adequacy rate were 90% (medium) and 101 % (good), respectively, while protein adequacy rate of household and individual levels were 85% (medium) and 94% (medium), respectively (unpublished data). This result was relevant to analysis of the cross classification between the proportion of food expenditure and the level of household energy consumption. With an average household income of IDR 1,000,000, per month and food expenditure of IDR 835,200, per month (Table 1), it can be said that a large part of household expenditure was allocated to fulfill basic needs (food).

Table 1. The average of food expenditure in the household (urban farming program in Sumberjaya 2018).

No	Food item	Expenditure (Rp 000)/month	% of food expenditure
1	Rice	302,0	36,2
2	Animal based side dishes	99,2	11,8
3	Plant based side dishes	88,0	10,5
4	Vegetables	150,0	18,0
5	Fruits	40,0	4,8
6	Vegetable oils	60,0	7,2
7	Legumes	40,0	4,8
8	Tubers	20,0	2,4
9	Sugar	36,0	4,3
	Total	835,2	100,0

*Data expressed as means (n=30).

Based on household food expenditures according to nine food items that were taken into account in determining the expected food pattern score (Table 1), 36% of the budget was still allocated to purchase the staple food, in the form of rice. The purchase allocation for animal based side dishes was around 11%, plant based side dishes was 10% and vegetables was 17%. The source of food was generally obtained through purchasing, except for rice harvested from their own field by those who had their own paddy field. In reviewing this action, it turned out that vegetable and plant based side dishes that were originally obtained from the market can be fulfilled from their own yard through the innovation of urban farming. The money spent for food can be saved up, thus improving their financial situation.

In the area of study, the consumption pattern of farmer households was dominated by rice, vegetables, eggs/fish and tofu/tempeh. Rice was the staple food, accompanied with mostly with vegetables whereas the popular types are kale and spinach. Egg was the most affordable menu of animal based protein and was the most easily cooked in addition to tofu or tempeh as a source of vegetable protein. Consumption of rice took the largest part of the household's food expenditure compared to other expenses. Though rice has a low protein content, this food was consumed in large quantities and frequently, therefore, rice contributed some parts in fulfilling protein requirement in a day (Adriani & Wirjatmadi 2012). Besides that, the protein requirement for the households was also achieved by the consumption of tofu, tempeh, eggs and fish, which were the largest source of protein besides meat and milk. After applying urban farming innovations in form of yard utilization, the community can improve the quality of consumption of vegetable and plant based foods in a self-sufficient way.

3.3 *Types of community adaptation towards ecological changes*

From the field observations, each participant was assigned their typology of ecology adaptiveness. We characterized four types of community adaptation to ecological changes and land use innovations, namely: (1) apathetic type, (2) reactive type, (3) proactive type, and (4) anticipatory type. This sequence also describes the more adaptive to changes that occur. Community profile based on the type of adaptation to the changes in the ecology in detail as explained bellow.

First, the apathy type subjects were characterized by no formal education or lower than elementary level of education, while their income level was low. This group was comprised of lower social class, so that their adaptation depended on the intervention or influence of other parties. This type was prone to follow the changes performed by the other parties, especially their neighbors or extension agents/field assistants or public figures/community leaders to persuade them to adopt the urban farming innovations.

Table 2. The correlation coefficient between education, activeness, contribution, initiative, performance, and adaptiveness.

Parameter	Correlation coefficient				
	Activeness	Contribution	Initiative	Performance	Adaptiveness
Education	0.307	0.419*	0.650*	0.659*	0.521*
Activeness		0.831*	0.761*	0.744*	0.725*
Contribution			0.700*	0.737*	0.798*
Initiative				0.933*	0.867*
Performance					0.837*

Data expressed as Spearman's rho coefficient correlation (n=30). Asterisk (*) indicated significance (two tailed) at $p<0.05$

Second, the reactive type subjects were characterized by the elementary level of education, the lower middle social class, and inclined to locality. This type adapted as a reaction to the difficulties that have been felt as well as the risk of the upcoming troubles, and through following their neighbors after being given information to apply innovation of the yard utilization.

Thirds, the proactive type subjects were characterized by people with secondary education, was relatively cosmopolitan and obtain information from assistants/counselors/figures outside their neighborhood. This type adapted by quickly taking action after receiving information from sources that can be accessed by both local leaders and figures outside the community. Usually these figures were people who became role models for other communities, and influenced other community members to implement innovation in facing changes in their environment.

Fourth, the anticipatory type subjects were characterized by people with a relatively high level of education and cosmopolitan, as well as access to digital cyber extension information and actively communicating with figures promoting the innovation (agents of change) such as counselors/assistants and relevant information from the internet. This type of subjects was able to digest information from various sources and design their adaptation by obtaining the knowledge about the impact of changes and risks that were expected to occur in their environment.

Based on data collected from observation by the field assistants, we calculated the correlations as shown in Table 2. We can see that the highest parameter correlated with adaptiveness was initiative ($r = 0.867$), followed by performance ($r = 0.837$). A correlation coefficient greater than 0.8 was considered strong, while a value less than 0.5 was considered weak (Bolboaca and Jantschi 2006). Thus, all the parameters observed were enough to be taken into account since their correlations to adaptiveness were above 0.5. Apart from their educational background, the initiative was not only highly correlated to adaptiveness but also to the performance level ($r = 0.933$).

Taking into account the correlation of education and adaptiveness ($r = 0.521$), we saw that even with quite low educational background of the participants, participatory assistance might be considered appropriate to improve the attitude of adaptation of society from apathy to anticipatory. Therefore, it is possible to minimize the risk of the ecological change impact by increasing the attitude of community adaptation. As has been described before, in relation to the threat of food security, it was observed that field assistance could increase the level of community adaptation towards the changes in urban farming.

3.4 *Strategy to develop urban farming innovation*

Reinforcing urban farming was effective and efficient when a business integration strategy for yard utilization was carried out both upstream and downstream accompanied by halal certification. In the upstream sector, which begins with strengthening the insights of land owners through comparative study activities and trainings conducted by PT. Pertamina EP in collaboration with CARE LPPM IPB. This activity was not only applicable as a capacity building tool, but also effectively serves as a medium for dissemination of information and innovative technology to the urban farming community. The increase in capacity means that it had been carried out in real terms in order to find innovations that support the competitiveness of urban farming products.

The provision of sustainable inputs was carried out by introducing environmentally friendly cultivation technologies, namely through the manufacture of organic fertilizers, procurement of nurseries as germplasm sources, and promoting the construction of rainwater reservoirs in community gardens. The empowerment became more effective after market development was carried out, both in local markets and through participation in exhibitions. Market expansion progressed widely in corresponding to product diversification. It was successful due to the actual support in increasing product competitiveness in the downstream integratively, namely through packaging, obtaining home industry product license and halal certification.

Urban farming could be an alternative model that effectively enhances household food security when it is developed based on local potential, appropriate technological innovations in terms of seeds, product packaging, group functioning and appropriate comparative study methods and community needs as market opportunities. This strengthens and enhances the strategy for developing food security that focuses on community empowerment, through increasing community independence and capacity to play an active role in realizing food availability, distribution and consumption over time (Purwaningsih 2008). Some recommendation to mitigate the impact of ecological change through innovation of urban farming were (1) participatory empowerment through synergetic cooperation of community, extension agent and field assistant, (2) integration of innovative upstream and downstream land businesses accompanied by halal certification, (3) dissemination of appropriate innovation to provide sustainable input, and (4) increasing the role of groups in product marketing partnership to improve farmer's bargaining position.

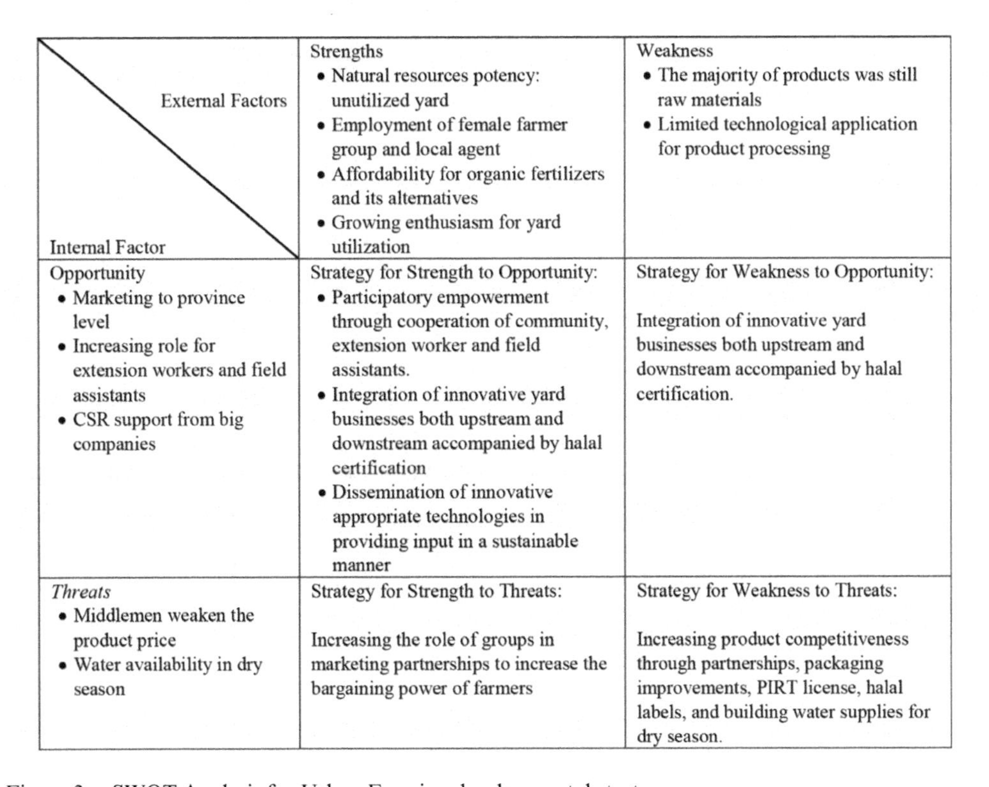

External Factors / Internal Factor	Strengths • Natural resources potency: unutilized yard • Employment of female farmer group and local agent • Affordability for organic fertilizers and its alternatives • Growing enthusiasm for yard utilization	Weakness • The majority of products was still raw materials • Limited technological application for product processing
Opportunity • Marketing to province level • Increasing role for extension workers and field assistants • CSR support from big companies	Strategy for Strength to Opportunity: • Participatory empowerment through cooperation of community, extension worker and field assistants. • Integration of innovative yard businesses both upstream and downstream accompanied by halal certification • Dissemination of innovative appropriate technologies in providing input in a sustainable manner	Strategy for Weakness to Opportunity: Integration of innovative yard businesses both upstream and downstream accompanied by halal certification.
Threats • Middlemen weaken the product price • Water availability in dry season	Strategy for Strength to Threats: Increasing the role of groups in marketing partnerships to increase the bargaining power of farmers	Strategy for Weakness to Threats: Increasing product competitiveness through partnerships, packaging improvements, PIRT license, halal labels, and building water supplies for dry season.

Figure 3. SWOT Analysis for Urban Farming developmental strategy.

4 CONCLUSION

The innovative use of the yard in encountering the food security threat was applied in developing urban farming in participatory way by employing appropriate technological innovations and considering the market for home business products. There were four types of community adaptation to ecological changes in the form of conversion of agricultural land to non-agriculture as a result of toll road construction and international airports: (1) apathetic type, (2) reactive type, (3) proactive type, and (4) anticipatory type. All stakeholders need to work together in encountering food security problem, in addition to controlling land conversion, it is also important to develop innovations in underutilized yard to be more productive. The innovation of yard utilization as a form of urban farming turned out to be an alternative solution to the threat of food security in the households.

ACKNOWLEDGMENT

The private sector of PT Pertamina EP had developed their concern for the villages directly affected by the construction of the Cipali toll road and Kalijati airport. It became a priority in their corporate social responsibility (CSR) program. We really appreciated their cooperation in developing this innovation in urban farming. This study also had received the fund from the Ministry of Research, Technology, and Higher Education of Indonesia. We would also like to thank all the extension workers and field assistants involved in the urban farming innovation as a part find alternative solutions for food security in conflict-prone villages.

REFERENCES

Adriani, M. & Wirjatmadi, B. 2012. *Introduction to Community Nutrition (Bahasa Indonesia)*. Jakarta: Kencana.

Ariani, M. & Rachman, H.P.S. 2003. Analysis of household food security level (Bahasa Indonesia). *Media Gizi dan Keluarga* 27(2): 1–6.

Bailkey, M. & Nasr, J. 2000. From brownfields to greenfields: Producing food in North American cities. *Community Food Security News* Fall 1999/Winter 2000:6.

Bolboaca, S. D., Jantschi, L. 2006. Pearson versus Spearman, Kendall's Tau correlation analysis on structure-activity relationship of biologically active compounds. *Leonardo Journal of Sciences* 9: 179–200.

Hidayat, Y., Ahyar Ismail, A., & Ekayani, M. 2017. Impact of agricultural land conversion to the economy of rice farmer households (case study of Kertajati sub-district, Majalengka district, West Java) (Bahasa Indonesia). *Jurnal Pengkajian dan Pengembangan Teknologi Pertanian* 20(2): 171–182.

Maxwell, S. & Frankenberger, T.R. 1992. Household food security: A Conceptual Review. Household Food Security: Concepts, Indicators, Measurements. A Technical Review. New York and Rome: IFAD/UNICEF.

Purwaningsih, Y. 2008. Food security: situations, problems, policies, and community empowerment (Bahasa Indonesia). *Jurnal Ekonomi Pembangunan* 9(1): 1–27.

Putri, N.P.A., Aini, N., & Heddy, Y.B.S. 2015. Sustainability evaluation of sustainable food house areas (*kawasan rumah pangan lestari* or KRPL) in Girimoyo Village, Karangploso District (Bahasa Indonesia). *Jurnal Produksi Tanaman* 3(4): 1–4.

Rahayu, A., Aziz, N., & Gagoek, H. 2013. Strategy for enhancing the sustainability status of Batu City as an agropolitan area (Bahasa Indonesia). *Jurnal Ekosains* 5(1): 21–33.

Supariasa, Bakri B., & Fajar. 2001. *Assessment of Nutritional Status* (Bahasa Indonesia). Jakarta: Penerbit Buku Kedokteran EGC.

Tindaon, F. 2015. Irrigation and conversion of paddy fields. https://www.researchgate.net/publication/280385643_Irigasi_dan_Konversi_Lahan_Sawah [Accessed at August 24, 2018].

Rural Socio-Economic Transformation – Kinseng et al. (Eds)
© 2019 Taylor & Francis Group, London, ISBN 978-0-367-23603-8

Agriculture and the regeneration problem: Rural youth, employment, and the future of farming

E. Soetarto, D. Nurdinawati, R. Sita, M. Sihaloho & T. Budiarto
Department of Communication and Community Development Sciences, Faculty of Human Ecology, Bogor Agricultural University, Bogor, West Java, Indonesia

ABSTRACT: This study aimed to describe the profile of farmer household in food farming sector and to analyze the agrarian structure of young farmers and opportunities for future farmers' regeneration. This study was conducted by comparing the conditions in Bantul Regency and North Bengkulu Regency, Indonesia. The results of this study indicated that the area of agricultural land controlled by young farmers in North Bengkulu is wider than in Bantul. However, in terms of the quality of its human resources, young farmers in Bantul showed better conditions, seen from many young farmers in Bantul who have high school education. This good quality of human resources enables young farmers in Bantul to be more adaptive to a narrow area of land and easy to adopt innovation, so they can meet their daily needs without having to make a double living.

1 INTRODUCTION

In agricultural development, farmers are the main actors who act both as subject and object. Farmers are the subject due to their position as economic actors in particular and citizens in general. Farmers are the main actors who determine the figure of the agricultural sector. On the other hand, in the context of evaluation or policy formulation, farmers (as well as other community groups) are objects whose characteristics need to be comprehensively understood. The ST 2013 results showed that the number of agricultural business households is currently 26.13 million households which decrease by 5.04 million house-holds compared to the results of the agricultural census in 2003. The loss of around 5 mil-lion farmers in the last ten years may be part of the transformation process of economic structures which marks the increasing level of progress of a country. In this case, the process is always marked by the continued decline of the agricultural sector's roles in the economy. While at the same time, the roles of the industrial sector and services are increasingly dominating. On the other hand, the worker reduction in agricultural sector could also be a bad sign, namely the existence of the tendency to exclude farmers from the agricultural sector which is usually an important source for them, so that in droves they begin to leave the work and choose to work in informal sector in the city.

When the development that occurs in Indonesia is urban-biased and does not favor the welfare of farmers in the countryside, the development imbalance between villages and cities that results in welfare inequality cannot be avoided. This leads to the phenomenon of urbanization in terms of the efforts of the poor to defend their lives after they have been excluded from the countryside. This phenomenon is different from the phenomenon of urbanization in the 1970s, where urbanization occurred was because the upper middle class villagers tried to "challenge new life" which considered to be more promising than if they remained in the village. The phenomenon of urbanization is also followed by the reduction of rural areas due to the rapid urban process over the past few years and this is a manifestation of urban-biased development. From a total of 75,410 villages in Indonesia, 24,716 villages have experienced urban processes and degradation of agricultural land. Agricultural land as a physical characteristic of the village has decreased by 100,000 ha/year (Ministry of Public Works 2012). Thus, statistics showed that the decline in the number of workers in the agricultural sector has been accompanied by a rapid rate of conversion of agricultural land to non-agricultural use. The reduction of agricultural land is due to the economic nature that develops in this country which is expansive capitalistic, so it is very "greedy" towards the use of space or land.

Another contributor to land conversion is the absence of incentives that can maintain the farmers' interests in their profession. As a result, they are often tempted to easily relinquish their agricultural land to the owners of capital so that this process contributes to the symptoms of land ownership reconciliation to a few people. In other words, they prefer to sell the land they own and the money is used as a capital to migrate to the city and work in the informal sectors. In addition, the land value which is increasingly expensive, coupled with the development tendency that continues to form an asymmetrical socio-agrarian relationship, make the weaker lose their land more quickly. In fact, for farmers, land is the most valuable asset. Its role is very important because it is a determinant of household in-come and often related to social status especially in developing countries (including Indonesia).

In the midst of the difficulty to live as a farmer in Indonesia, there are still 26.13 million included in agricultural households. They are the foundation to meet the food needs of 250 million people in this country. To them the development of agriculture depends. One of the important actors in agriculture who also determines the future of agriculture in Indonesia is the young farmers. If agriculture no longer promises welfare, the reluctance to work in agricultural sector will increasingly be visible. Easier access to education even to the rural level makes these young people far more "educated" than youth in the past, meaning that the employment choices will be more diverse for them. Therefore, there is a concern about the future of agriculture in Indonesia.

Ironically, being a farmer is not idealism for young people especially in developing countries (White 2011, Leavy & Hossain 2014). Fewer young people are interested in becoming farmers. Agriculture is seen as the domain of the elderly. Young people tend to think that the profile of young people is dynamic, only they can achieve it in the more profitable non-agricultural sectors. Working on agriculture is still seen as a difficult job and is not financially profitable. Agriculture is seen as a job with a low status, and dirty (low status, dirty work). Regeneration of farmers is a crucial problem because the sustainability of the agricultural sector in the future depends heavily on the gait of these young people. Young people should be the future of the agricultural sector that replaces the aging population of farmers. The strength of young people in the agricultural sector lies in energy, capacity and ability to produce ideas and innovation (Chinsinga & Chasukwa 2012). However, there are some young people who still want to work as farmers, so we want to know: (1) what are the social, economic and demographic character-istics of young farmers in rural areas in Java and outside Java? Do they really choose to be farmers or be excluded because they cannot work in the sector they want?; 92) what are the structure and direction of changes in agricultural land ownership as well as their changes' direction and implications for farmers in particular and rural households in general?; and (3) what impact do young farmers feel and what are their attitudes towards rural agriculture?

Halamska (2011) told that the reduction in the number of farmers in many regions in Indonesia, including in rural areas of Java and outside Java, resulted in a depreciation process, which would then have an impact on the reduction of rural areas (deruralization). As an agricultural country, this will certainly threaten the process of deagrarianization in the future. As explained in the previous section, the phenomenon of the reduction in the number of farmers in a country may be part of the transformation process of economic structures which marks the increase of the progress level of a country, but on the other hand, it can also indicate the tendency of farmers to be excluded from the agricultural sec-tor which is usually an important source for them, so they leave the job in droves. Thus, there is a sense of curiosity about "Why has the agricultural sector in Indonesia increasingly lost its workers? Is this a good sign which means that there is an economic transformation to a better direction or is this a bad sign which means that there is an urban-biased development, resulting in poverty in rural farmers? Are farmers eliminated from their livelihoods? Is this a sign of depreciation?"

2 LITERATURE REVIEW

The term 'young people' can be seen as a social category or a biological category. They can be included in category of action, (sub)culture practice, identity, and generation. In official vocabulary, the term 'youth' is used not to mark age or biological construction, but it is often used for social construction. The use of the term 'young' often functions to make them as a transitional period in which in power relations it functions as a safety zone for the position above it, and its existence can be ignored at any time when it is not need-ed. Ben White (2011) stated, "to exclude them from mainstream social, economic and political processes as something less than full members of society, less than full citizens". This term was defined in the observations of parents who saw it as an unstable phase. In fact, there is a tendency for the state (a substitute for patrons as parents) to extend the age of the designation of young people for this purpose (for Indonesia 18-35, Malaysia 18-40/45).

The crisis of regeneration becomes a concern for the future of agriculture. In the span of almost a decade, the rate of employment in the agricultural sector in Java declined, from 43% (in 1993) to 32% (in 2010). Data from BPS in 2011 stated that only around 6.9 million or around 11% of 62.92 million youths who are working in the agricultural sector. The rest of them work in other sectors that cannot also be defined as permanent workers. The decline in the younger generation of agriculture is in line with the total decline of agricultural labors in Indonesia. The results of the Agricultural Census of BPS in 2013 in-formed that in the span of the past decade there had been a decline in the number of agricultural households (RTUP) of food crops by 979,867 people from 18,708,052 people (2003) to 17,728,185 people (2013). If it is coupled with non-food crops RTUP, the total number of farmers in Indonesia was 26.13 million (BPS 2013) which also experienced a decline in numbers.

The lower number of youth in land-based works in various countries in the world is a result of complex land policies and structural conditions (Ben White 2011). Ben White explained that there are several factors that cause the lower proportion of young people working in the agricultural sector: (1) decline in knowledge and expertise in the field of agriculture (deskilling youth on agriculture knowledge); (2) decline in agricultural and rural life due to urban-biased development and policies; (3) even if young people want to farm, there is no access to land for them. This is related to the patriarchal culture and gerontocracy, where people are oriented to the importance of adults rather than children; (4) there is a serious problem that occurs in Indonesian rural families, namely the re-lease of family land. Parents capitalize on children's education by selling their reserves of wealth, especially land; and (5) threats of extractive economic development and infra-structure development that convert agricultural productive lands, or change household scale agriculture into a corporate scale.

In the future, young people who are still surviving as young farmers will face challenges that are also not easy. These challenges include: (1) lack of access and control over pro-ductive resources, especially land and capital; (2) low skills and knowledge about business production, processing and management; and (3) globalization, uncertainty, price diversi-ty

The aging of the average age of farmers is a serious demographic phenomenon in agriculture. Structural changes in the employment demography of the agricultural sector also occur in Asian and European countries, as well as in America, Canada, and other continents. This shows that the decline in interest in agricultural sector has become a common phenomenon that needs serious attention from policy makers in order to save the agricultural sector. The results of the Agricultural Census in 2013 showed that the proportion of farmers with more than 40-54 years of age was the largest, namely 41%. The second largest proportion was the age group of more than 55 years which can be classified as an old farmer, which was 27%, while the younger generation group with the age of 35 years was only 11%. The Agricultural Census in 2003 also showed that the majority of farmers was in the age group of 25-44 years (44.7%), then followed by the age group of 45–60 years by 23.2%. The proportion of the elder labors (> 60 years) was around 13.8%, and the lowest was the young one (< 24 years) which was only 9.2%.

Susilowati (2016) explained that the development of agricultural employment as outlined above strengthens the phenomenon that young rural workers tend not to choose agriculture as their job. They tend to go to the cities to find work in other sectors. The decision of these young workforces is mainly due to the motivating factors, including agricultural land which is increasingly narrow and not economical to cultivate. From the economic point of view, their decision to look for jobs outside the agricultural sector is rational, considering that the agricultural sector is seen as unable to meet the needs of life. Young workers who have just started a business in the agricultural sector have limited financial capacity to own large areas of land, unless they obtain inheritance or work on the property of their parents. With a land tenure of less than 0.25 ha, it is very unattractive for young farmers to start doing business on land-based agriculture or conventional farming (food crop farming).

3 METHODOLOGY

3.1 *Description of the study area*

The research is planned to be conducted within two years, namely in 2018 and 2019. The first-year research was conducted from July to August 2018, aiming to obtain base-line data on the profile of young farmers in the research locations, which then the base-line results are used as the basis for the second-year action research formulation. This first-year research took places in Sidomulyo Village, Bambanglipuro Subdistrict, Bantul Regency, and in Tebing Kaning Village, Armajaya Subdistrict, North Bengkulu Regency. The selection of these two locations was based on the initial purpose of the study, namely to study the diversity of social, economic, and demographic characteristics of farmers in rural areas in Java and outside Java, especially young farmers on food agriculture. The two locations were chosen as study areas because the agricultural sector in both areas survives although other sectors come in. Bantul is invaded by the tourism services sector, while Bengkulu is devastated by the oil palm plantations. In the midst of the onslaught, the food crop farm-ing in these two villages in Bantul and Bengkulu still survive, in where the farms are still being worked on by young farmers. This is seen from the large proportion of the main farmers under the age of 45 years in the two regencies compared to others.

3.2 *Research design*

This research was designed using quantitative and qualitative approaches. Quantitative approach was carried out with a survey, namely data collection using questionnaires, while qualitative approach was done by Focus Group Discussion (FGD), in-depth interviews, as well as observations on social, economic, cultural, and environmental conditions of the study locations. The subject of this research consisted of respondents and informants. A total of 40 respondents were selected at each location, namely young people involved in the food sector of paddy agriculture, both as land owners, cultivators, and farm workers under the age of 45 years. The age of 45 years was chosen because this age is still under the median of food crop farmers in Indonesia which is 47 years. The informants in this study consisted of government agencies such as the agricultural service, Development Planning Agency at Sub-National Level (BAPPEDA), agricultural extension officers, village and sub-district governments, agricultural leaders in the village, and youths.

3.3 *Data collection strategies and instruments used*

According to the data source, this research data consisted of primary data and secondary data. Primary data were obtained directly in the field by structured interviews using questionnaires, in-depth interviews with informants, observations, and focus group discussion (FGD). Meanwhile, secondary data were obtained by utilizing available data such as the publication of BPS data on agriculture, regencies in numbers, village monograph data, data in agencies visited, and others. Some data collected were quantitative, while some others were qualitative.

Questionnaires, in-depth interview guides, and FGD guidelines were some examples of the data collection instruments used in this study. The questionnaire consisted of several sections that were quite comprehensive so that they could be used to explore data optimally. The questionnaire covered individual and respondent's household characteristics, agricultural land tenure, the needs of young farmers in developing agriculture, livelihood structure, capital ownership, and perceptions of agriculture.

3.4 *Data analysis and presentation*

The quantitative data that had been obtained were then processed and analyzed descriptively and inferentially. Descriptive analysis was presented in the form of frequency tables and images. The inferential analysis was presented in the form of cross tabulations, for example to see the relationship between individual and household characteristics with the pattern of land tenure. The analysis process of qualitative data was done by coding (categorization) data from library research, in-depth interviews and case studies, based on specific issues/topics that will answer specific research questions. For the needs of data analysis, interview guidelines were made in the form of tables with the main themes and specific questions that could answer certain problems from the perspective of different respondents and informants. In addition, analysis of the contents of mass media articles and/or written thoughts and opinions was also carried out. Qualitative data obtained from these various methods of data collection were then analyzed from the initial stage of re-search according to three lines of activities carried out simultaneously, namely: data reduction, data presentation, and data conclusion or verification (Miles & Hubermen 1992).

4 RESULTS AND DISCUSSIONS

4.1 *Social, economic, and demographic characteristic of youth farmer*

Table 1 shows the general characteristics of young farmers in both locations used as samples. As explained in the methodology, the number of samples from each regency was 40 people. The general description of the characteristics of the respondents informed the distribution of respondents in both locations according to gender, age, education level, marital status, household status, household income level, status of controlled paddy fields, and number of years of experience as a farmer. Descriptive statistics about these variables are displayed in the form of a percentage of the total respondents for each regencies (40 people) or 80 people for the total respondents as a whole.

Based on Table 1, it is known that there were similarities in individual and household characteristics between respondents in Bantul Regency and North Bengkulu Regency regarding the type of gender, age, marital status, household status, and control status of paddy fields. It appears that in the two regencies the majority of respondents was young male farmers, with an average age of 36 years, already married, being the head of the house-hold, and having status as cultivators in the paddy fields they controlled. The differences between these two locations lies in the variables of level of education, average income, and number of years of experience as a farmer. Young farmer respondents in Bantul Regency showed better conditions of education and income levels compared to those in North Bengkulu Regency, but they had a shorter number of years of experience as a farmers compared to young farmer respondents in North Bengkulu Regency. Young farmers in Bantul had a higher level of education than farmers in North Bengkulu. With a higher level of education young farmers in Bantul had experience in working outside the agricultural sector but then choosing to go back to the agricultural sector. It is different from young farmers in North Bengkulu who had chosen the agricultural sector as their source of income since the beginning. Since agriculture has become an option since the beginning, young farmers in North Bengkulu had more experience in farming than young farmers in Bantul who had chosen alternative economic sources outside of agriculture.

4.2 *Agrarian structure of youth farmer household*

Table 2 shows the pattern of agricultural land tenure based on the characteristics of young farmers in both locations. The land tenure of farmers in Bantul is narrower compared to farmers in North Bengkulu. On average, a total of 86 percent of the land holdings by Ban-tul farmers were below 0.24 hectares. This means that young farmers in the village of Bantul were farmers with very narrow land. Meanwhile, a total of 52 percent of young farmers in North Bengkulu controlled the land on average over 0.5 hectares.

Table 2 shows that on a variety of individual characteristics, both based on education level, age and marital status, the majority of young farmers' land tenure was below 0.25 ha. This means that these individual characteristics did not affect the patterns of land ownership by

Table 1. Main characteristics of youth farmers.

Characteristics	Bantul (%)	North Bengkulu (%)	Total (%)
Sex			
1. Male	91	95	93
2. Female	9	5	7
Age			
1. 21-25	14	8	11
2. 26-30	9	15	12
3. 31-35	30	25	28
4. 36-40	19	23	21
5. 41-45	28	30	29
Education level			
1. Don't go to school	0	5	2
2. Elementary school	5	33	18
3. Junior high school	21	38	29
4. Senior high school	67	23	46
5. University	7	2	5
Marital status			
1. Single	23	13	18
2. Married	74	88	81
3. Divorced	2	0	0
Relationship status with Household head			
1. Household head	65.1	85	74.7
2. Wife	9.3	5	7.2
3. Son/daughter	25.6	10	18.1
Household income (IDR/year)			
1. Low (≤14,288,535)	19	27	23
2. Medium (14,288,535-41,148,958)	65	60	63
3. High (≥51,148,958)	16	13	14
Land ownership status			
1. Owner and cultivator	34.9	37.5	36.1
2. Cultivator only	65.1	62.5	63.9
Years of farming experience			
1. ≤7	63	18	42
2. 7-14	21	26	23
3. ≥14	16	56	35
Average number			
1. Age	35.6	36.3	35.9
2. Household income (IDR/year)	38,040,835	26,997,500	32,718,745
3. Years of farming exprience	7	14	11

young farmers. Generally, the land areas controlled by farmers in Bantul was strongly influenced by the ownership factor of the parents' land. The land cultivated by young farmers came from the inheritance of parents. The same thing was applied for young farmers in North Bengkulu who controlled the land from the giving of parents. Therefore, land tenure between generations over time will be increasingly narrow due to the segmentation of land following the number of children owned. The findings in the fields, especially in Bantul Regency, showed that even though overall agricultural land area was relatively fixed, individual ownership/household ownership was narrowing over time. Meanwhile, with the increase of agricultural land scarcity, it was increasingly difficult for farmers to add more agricultural land.

4.3 *Agriculture as a livelihood strategy*

The young farmers controlling the agricultural land as shown in Table 2 did not fully have the status of an owner, but some of them were also cultivators, both working on land owned by parents and by other people or family relatives who run a sharing system. With the condition of the

Table 2. Percentage of young farmers based on the land area controlled and the individual characteristics.

Characteristics	Bantul			North Bengkulu		
	<0.24 Ha	0,25-0,49 Ha	>0,5 Ha	<0.24 Ha	0,25-0,49 Ha	>0,5 Ha
Percentage	86	14	0	8	40	52
Sex						
1. Male	85	15	0	5	40	55
2. Female	100	0	0	50	50	0
Age						
1. 21-25	83	17	0	0	33	67
2. 26-30	100	0	0	0	50	50
3. 31-35	91	9	0	20	50	30
4. 36-40	86	14	0	0	33	67
5. 41-45	82	18	0	8	33	58
Education level						
1. Don't go to school	0	0	0	0	100	0
2. Elementary school	100	0	0	15	39	46
3. Junior high school	100	0	0	0	47	53
4. Senior high school	80	20	0	0	22	78
5. University	100	0	0	100	0	0
Marital status						
1. Single	80	20	0	0	40	60
2. Married	88	12	0	9	40	51
3. Divorced	100	0	0	0	0	0
Years of farming experience						
1. ≤7	85	15	0	14	43	43
2. 7-14	100	0	0	10	40	50
3. ≥14	71	29	0	5	36	59

narrow land, young farmers in Bantul were more adaptive to horticul-tural agriculture. Agriculture with a rotation system between food agriculture and horticul-tural agriculture on narrow land was the choice of young farmers to survive in the agricul-tural sector. With a rotation system, 52 percent of young farmers relied solely on agricul-ture as a source of income. The remaining 46 percent of farmers, in addition to farming, also did other works outside the agricultural sector such as by becoming factory workers and unskilled laborers. Meanwhile, with wider land tenure the pressure for expansion of rubber and oil palm plantations was a challenge for agriculture in North Bengkulu. In addi-tion to working on agricultural land, as many as 50 percent of respondents of young farm-ers in North Bengkul worked as laborers for oil palm and rubber plantations.

Table 3 shows that the majority of young farmers both in Bantul and North Bengkulu had jobs outside the agricultural sector in the increasingly higher age groups. Meanwhile, at various levels of education respondents in both locations had the same tendency. Some only did farming and some had jobs other than farming. According to the marital status, unmarried young farmers in Bantul partially chose to only work in the agricultural sector, while those who were married also had other jobs outside the agricultural sector. Unlike the respondents in North Bengkulu, the majority of those who were unmarried had more jobs outside the agricultural sector than those who were married. It is interesting that even though the size of the controlled land by young farmers in Bantul was narrower compared to those in North Bengkulu, more young farmers in Bantul chose to work only in the agri-cultural sector compared to those in North Bengkulu who preferred to have other jobs out-side the agricultural sector .

Young farmers in Bantul Regency made agriculture as the main sector of their livelihood. Even farmers who plunged in other sectors chose to return to work in the paddy agricul-tural sector. However, the young farmers in Bantul felt that they still need to be contribut-ed by the sectors outside of agriculture because it is not enough to only depend on the ag-ricultural

Table 3. Percentage of young farmers based on the livelihood strategies and the individual characteristics.

	Bantul		North Bengkulu	
Characteristics	Only farming	Have another job	Only farming	Have another job
Percentage	54	46	50	50
Sex				
1. Male	54	46	50	50
2. Female	50	50	50	50
Age				
1. 21-25	100	0	67	33
2. 26-30	50	50	50	50
3. 31-35	38	62	20	80
4. 36-40	62	38	56	44
5. 41-45	42	58	67	33
Education level				
1. Don't go to school	0	0	0	100
2. Elementary school	50	50	54	46
3. Junior high school	56	44	53	47
4. Senior high school	55	45	56	44
5. University	33	67	0	100
Marital status				
1. Single	80	20	40	60
2. Married	44	56	51	49
3. Divorced	100	0	0	0
Land ownership status				
1. Owner and cultivator	53	47	53	47
2. Cultivator	54	46	48	52
Land are controlled				
1. $\leq$ 0.24 Ha	58	42	33	67
2. 0.25-0.49 Ha	80	20	50	50
3. $\geq$0.5 Ha	0	0	52	48

sector with the current limited land area. Meanwhile, young farmers in North Bengkulu, mostly made the agricultural sector as a side job. The main works for them were as oil palm workers, rubber workers, and workshop workers. Only a small percentage of young farmers made the agricultural sector as the main job.

5 CONCLUSION AND RECOMMENDATION

Agriculture is a strategic sector in ensuring the sustainability of development and society (Luckey 2013). With regards to this, the way to attract young people and to hold them to be willing to continue to work in the agricultural sector still seems to be a problem and a global challenge. Developing countries, including Indonesia, still face the same complicat-ed problems. Another problem concerns the question of how to ensure food security in the midst of relatively high population growth, but parallel to that there is the declining involvement of young people in the agricultural sector.

It is a relief when realizing that employment opportunities in the agricultural sector in various parts of the world are reported to be still available and even in relatively large num-bers and tend to increase over time, including the potential employment of agricultural graduates. But it is also realized that the younger generation today is more inclined to be reluctant to choose the agricultural sector as a profession and job career (Mukembo 2014). Whereas, youth involvement in agricultural activities is a key factor in the sustainability of agricultural development because youth is a potential generation with a character that tends to resemble innovators and fast learners. Its young age is an important asset in ensuring the sustainability of agricultural development

(Kimaro et al. 2015). This is where returning or inviting young people back to agriculture becomes a challenge, and this largely depends on the ability of the parties to bring up attractive factors that may be able to invite young people to work in the agricultural activities.

In the Indonesian context, especially in Java and Outside Java, empirical findings indicate that the role of young farmers is still stretching. Incentives, creations and innovations are still very likely to give to support their existence and action to ensure the sustainability of the agricultural relay. But before that there are a number of questions: how far the decision makers in particular are able to produce a comprehensive blue print of agricultural (and rural) development which takes side to the interests of farmers. How far the stake-holders in this regard are committed and endeavor to consistently correct and overcome the structural problems that accompany the problematic agricultural (and rural) sectors, such as the issue of criticism towards agrarian resource arrangements, poverty, unemployment and other socio-economic gaps that must be executed with affirmative policy instruments that are optimal and address the roots of the problems. Returning young farmers is a necessity as part of the village's safeguards and the welfare of its people so as the villages do not sink deeper into the process of deruralization, depeasantion and deagrarianization (Soetarto 2010).

One of the fundamental problems of the agricultural transition in Indonesia that arises from the current pro-growth policies is the issue of the gap in the control of agricultural lands. The issue of 'landless' is part of crisis that must be addressed immediately to support young farmers who are interested and willing to strengthen their positions in agricultural activities in the countryside. Agriculture that is already running and considered still capable of attracting young farmers in fact is still gripped by the uncertainty of the availability and control of current and future agricultural lands. Whereas, the guarantee of sufficient land tenure for farmers is a basic precondition to answer the issue of farmers' regeneration discontinuation. This precondition must really become special concern because young farmers who currently survive and take part in the agricultural sector have never escaped the threat of various other massive non-agricultural expansions in taking over the control and designation of productive agricultural lands. If the solutions of the problems on ownership or tenure of agricultural land at a minimal scale for farmers cannot be found, the wave of transition of young farmers to the non-agricultural employment sector will be a necessity in the present and in the future.

REFERENCES

Chinsinga, B. & Chasukwa, M. 2012. Youth, Agriculture and Land Grabs in Malawi. *IDS Buletin* 43 (6):66-77.

Halamska, M. 2011. The Polish Countryside in the Process of Transformation 1989-2009. *Polish Sociological Review* I(173):35-54.

Kimaro, P.J., Towo, N.N., & Moshi, B.H. 2015. Determinants of Rural Youth's Participation in Agricultultural Activities: The Case of Kahe East Ward in Moshi Sural District, Tanzania. *International Journal of Economics, Commerce and Management* 3(2):1-47.

Leavy, J. & Hossain, N. 2014. Who Wants to farms? Youth Aspirations, Opportunities and Rising Food Prices. *Institute of Development Studies (IDS) Working Paper 439.*

Luckey, A.N., Murphrey, T.P., Cummins, R.L., & Edwards, M.B. 2013. Assessing Youth Perceptions and Knowledge of Agriculture: The Impact of Participating in an Agventure Program. *Journal of Extension* 51(3):1-7.

Miles, M. B. & Huberman, M. 1992. *Analisis Data Kualitatif.* Jakarta: Penerbit Universitas Indonesia.

Mukembo, S.C., Edwards, M.C., Ramsey, J.W., & Henneberry, S.R. 2014. Attracting Youth to Agriculture: The Career Interests of Young Farmers Club Members in Uganda. *Journal of Agricultural Education* 55(5):155-172.

Soetarto, E. 2010. *Desa dan Kesejahteraan Rakyat: Menegaskan Hak-Hak Dasar RakyatSebagai Platform Pembangunan, dalam Pembangunan Perdesaan dalam Rangka Peningkatan Kesejahteraan. Pemikiran Guru Besar PT BHMN.* Bogor: IPB Press.

Susilowati, S.H. 2016. Fenomena Penuaan Petani Dan Berkurangnya Tenaga Kerja Muda Serta Implikasinya Bagi Kebijakan Pembangunan Pertanian. *Forum Penelitian Agro Ekonomi* 34(1):35-55.

White, B. 2011, 'Who Will Own the Countryside: Dispossession, Rural Youth and the Future of Farming', *Valedictory Lecture, ISS*, 13 October 2011.

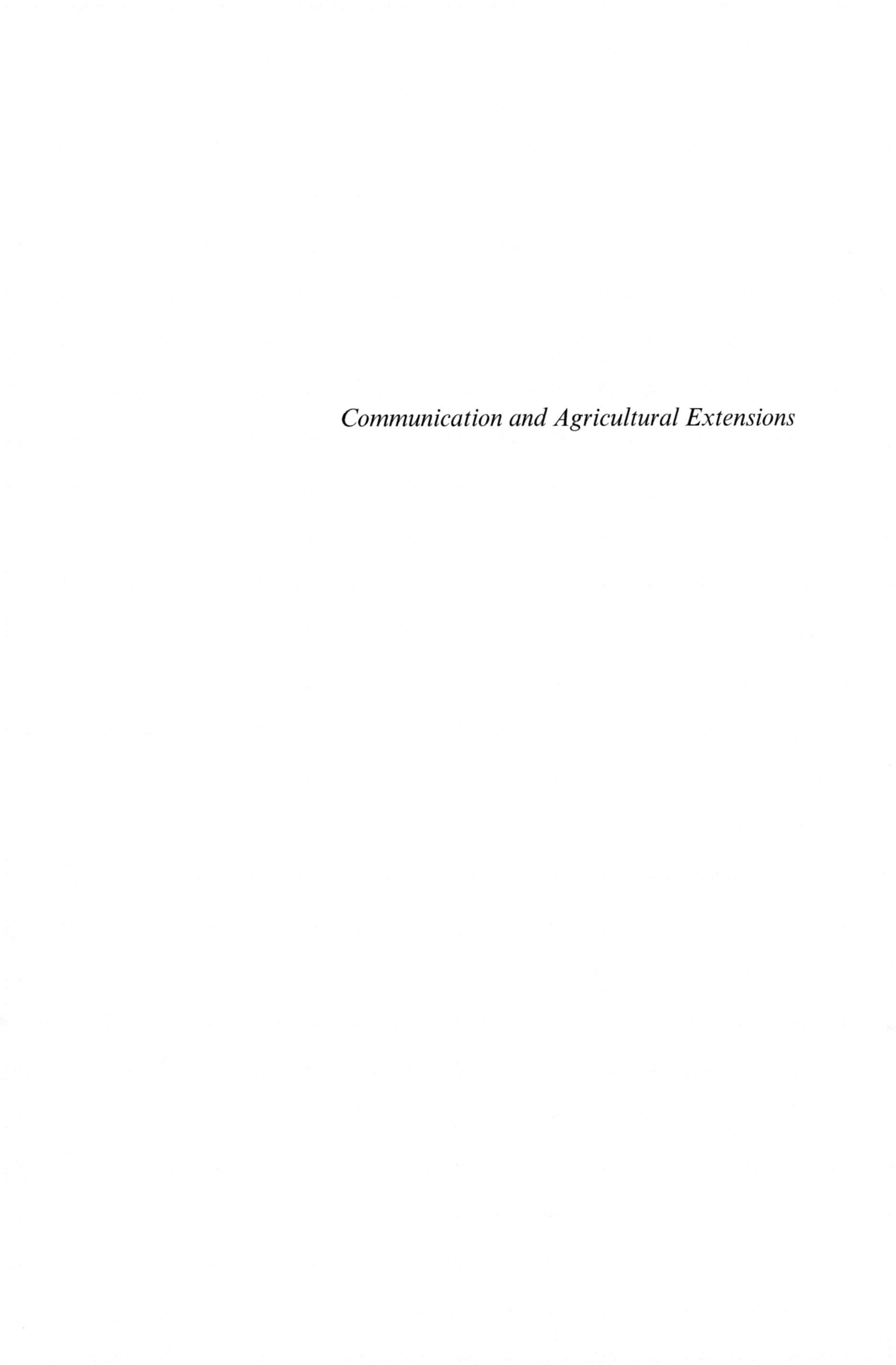

Communication and Agricultural Extensions

Rural Socio-Economic Transformation – Kinseng et al. (Eds)
© 2019 Taylor & Francis Group, London, ISBN 978-0-367-23603-8

The strategies to improve the sustainability of fish processing business through extension services and fish processing innovation

A. Fatchiya, S. Amanah & T. Soedewo
Department of Communication and Community Development Sciences, Faculty of Human Ecology, Bogor Agricultural University, Bogor, West Java, Indonesia

ABSTRACT: This research aimed to measure the performance level of innovation provision and the performance level of extension services, as well as to set up the strategies to increase the sustainability of fish processing business. Data were collected by survey with samples of 80 persons in Cirebon Regency, West Java Province. Data were collected in April 2018 and analyzed using Importance Performance Analysis, Customer Satisfaction Index, and a matrix. The results showed that the fish processors considered the performance of extension services and innovation to be in the category "achieved", except for extension services on business capital facility and product packaging innovation. The strategies to improve the sustainability of fish processing business for extension services were focused on the role of extension workers to obtain business capital, while for fish processing innovation the strategies were focused on the ways to obtain business capital and the use of new packaging.

1 INTRODUCTION

The fish processing business actors are one of the agents in a fishery-business chain system. Fish processors function to process fish raw materials to become processed foods ready to eat. They are also marketers or fish traders that become the distributors of fishery products to the consumers.

In general fish processing in developing countries is carried out by small-scale fish processors; for example, in African countries 85 percent is carried out by artisanal fisheries (FAO 1992). Likewise, in Indonesia 98 percent of fish processing is carried out by small and micro-scale business, and the rest 2 percent by middle and big-scale business (Ministry of Marine and Fisheries 2014). In Cirebon Regency, the fish processing center in West Java, 98% of fish processing is carried out by small and micro-scale business (Fishery and Marine Office, Cirebon Regency 2015), and it is managed traditionally.

Some products of traditional processed fish are dried fish, boiled fish, fish cackers, and smoked fish. The characteristics of artisanal fisheries in many areas in Indonesia are in general they do not have business license from the government. The workers come from unpaid family members. Their product is very limited and occasional, and their financial management is not recorded (Suwardane 2018). They are not capable of handling raw materials (Devi et al. 2016), and they use simple technology (Lumban et al. 2014). They pay less attention to the hygienity of raw material (Junianingsih 2014). They are less innovative, only relying on generation-to-generation techniques, and they do not apply the Hazard Analysis and Critical Control point (HACCP). They also have limited access to resources, such as capital, market, training, and information (Yanfika et al. 2018).

The limited business carried out by fish processors needs to be solved so that it can become sustainable and more developed, and in turn it can improve their welfare. For this reason, it is necessary to have a comprehensive and thorough approach system, namely extension. The community development programs for fish processors in Indonesia are carried out through fishery extension under the responsibility of the Ministry of Maritime and Fisheries. Fishery extension in Indonesia has proven to be able to increase the fish processors' welfare (Hudaya et al. 2018) and capacities (Yanfika et al. 2018, Suwardane et al. 2018).

To reach the sustainability of fish processing business, besides extension services, there needs also an innovation. Some research results showed that technological innovation that is applied on the traditional processing has been able to increase product quality, such as smoking innovation (Nti et al. 2002, Kumolu-Johnson et al. 2010, Salan et al. 2006, Esterlina & Nanlohy 2016) and can be stored under ambient room conditions (Holma et al. 2013).

Innovation means something new. According to Rogers (2003), van den Ban & Hawkins (1996), and Leeuwis (2009), something new does not only refer to technology, but more than that, namely a social and collective dimension, such as community group control, marketing, and new forms of interaction. Therefore, in this research innovation does not merely involve processing technology, but also the following aspects: market, equipment, storing and packaging, capital and others. According to Tutuarima (2016) the quality of processed fish product is influenced by the processing, packaging, transportation, storage conditions, as well as the manner of presentation for sale in the market. According to Njai (2000) problems faced by traditional fish processors are low quality of raw fish material due to poor fish handling in the artisanal sector, improper packaging of raw material, inappropriate processing procedures which result in low quality, and poor packaging technique so that the product is non-durable and contaminated. In terms of market aspect, there is no transparent market information, in spite of the fact that it is a vital factor for efficient production planning. In terms of transportation aspect, there are poor transportation facilities, so the product distribution to consumers cannot flow smoothly.

It is necessary to have an appropriate strategy to develop sustainable traditional fish processing in accordance with the problems faced. According to Njai (2000) there needs to be a collaboration between the governemnt and fish processors to build facilities for storing fish and to strenghen extension services through training and provision of two-way communication forum. The government also needs to formulate a policy to recover fish industry, and to synergize laws, regulations and development of the strategy implementation. Formal and non-formal education needs to be improved with regards to the importance of the resource and also the importance of improved processing, storage and distribution. Yanfika et al. (2018), Hudaya et al. (2018), and Suwardane et al. (2018) suggest that it be important to improve the capacity of traditional fish processors through intensive extension, so that it can increase their welfare.

Therefore, the aims of the research are as follows: (1) to identify the extent to which the performance of innovation and extension services meet the fish processor community's expectation, and (2) to make strategies to improve the sustainability of fish processing business through extension services and innovation of fish processing technology based on the fish processors' performance level and expectation.

2 METHODS

The research location is in Cirebon Regency, West Java Province, Indonesia. This location has been chosen because it is the center of fish processing industry. Three districts of fish processing center have been chosen from this regency. The research was carried out for four months, from April to August 2017. The research approach used mixed-method: first using a quantitative approach through survey; and second using a qualitative approach, i.e. Focused Group Discussion (FGD) with 15 participants from fish proccesor groups. This FGD was carried out to get further information about needs priorities, innovation performance and extension services. FGD was also carried out to get the strategic formula for improving business sustainability from the viewpoint of the fish processors themselves. The number of population was 348 from Marine and Fisheries Office of Cirebon Regency (2016), consisting of 135 persons from Gunungjati District, 117 persons from Suranenggala District and 96 persons from Jamblang District. The number of samples was 78 persons, determined using Slovin formula, with the error tolerant limit up to 10 percent. Samples were taken using a stratified random technique based on each district category. Slovin formula (Riduwan 2012) is as follows:

$$n = N/(Nd^2 + 1) \quad (1)$$

Where n=number of samples, N=number of population, d^2=precision (determined 10% wit a 90% trust level).

The number of samples for each district is determined by a proportional technique which refers to the following formula:

$$ni = Ni/Nxn \quad (2)$$

Where ni =number of samples according to stratum, n=number of total samples, Ni=number of population according to stratum, and N=number of total population.

Based on the above sample determination according to strata, details of population number and samples are presented in Table 1.

The instrument used was a questionnaire using closed questions with four answer choices of the respondents' agreement (modified Likert scale) for 10 variables (attributes), each for extension services and the availability of innovation. The attributes were classified into two groups according to their importance and performance level. The importance level was rated through the respondents' perspective on the priorities of extension services and innovation needed, and the performance level was rated from the respondents' perspective on the achievement level of the extension workers and the availabilty of innovation. Data completion in the questionnaire was carried out by enumerators based on the interview with each respondent.

The data analysis used to answer the first objective was by using the Importance Performance Analysis (IPA) grid and Customer Satisfaction Index (CSI) grid. IPA grid is useful to rate the customer satisfaction and to make management strategies (Sever 2015). IPA grid can also be used to rate the quality of services in various sectors, such as tourism (Wong et al. 2011). The IPA formula is as follows (Supranto 2011):

$$Tki = (Xi/Yi) \times 100\% \quad (3)$$

Where Tki=respondent suitability level, Xi=performance evaluation score and Yi=importance evaluation score.

Evaluation on each attribute in performance (x) and importance (Y) is 1 to 4 (Table 2). To measure the satisfaction level, there are four criteria used (Table 3).

The analysis to get the answer for the second objective was by using cartesius diagram, which consists of four quadrants: I, II, III, and IV. The position of each attribute in each

Table 1. Number of respondent distribution.

District	Number of population	Number of samples
Gunung Jati	135	30
Suranenggala	117	26
Jamblang	96	22
Total	348	78

Table 2. Importance and performance level based on weight.

Weight	Importance Level (Y)	Performance Level (X)
1	Unimportant	Unsatisfied
2	Less important	Less satisfied
3	Important	Satifsfied
4	Very important	Very satisfied

Table 3. CSI score and criteria.

CSI score	CSI criteria
0.76-1.00	Satisfied
0.51-0.75	Quite satisfied
0.26-0.50	Less satisfied
0.00-0.25	Unsatisfied

quadrant was calculated according to the average weight. The average weight of performance evaluation (X) showed the position of an attribute on the axis (X), whereas the position of attribute on the axis (Y) was shown by the average weight of the respondent's importance level (Y). Quadrant I shows the strategy as the main priority, quadrant II as maintaining the good work, quadrant III as low priority, and quadrant IV as an exessive strategy.

3 RESULTS AND DISCUSSIONS

Innovation in the fish processing business management in this research is not only from the novel technology and production techniques, but also from the capital and marketing aspects. The most needed innovation by the fish processors is business capital with low interest from a financial institution (Table 4). So far the main capital of the traditional fish processors has come from themselves or has been borrowed from middlemen and a "travelling bank" with very high interest. The lowest suitability rate between performance and innovation needs is the capital attribute (Table 4). This shows that the performance of financial institution or bank has not met the fish processors' needs for business capital. Nevertheless, in total the satisfaction level of the fish processors on the availability of innovation shows "quite satisfied" with score 63,75 on range score 57-75 (Table 4). This means that in general respondents feel satisfied, except for the existing innovation. This happens because their expectation for the needed innovation (importance) is higher, that is 32.62, compared to the existing innovation (performance) which is 25.29.

Based on the fish processors' perspective, the kind of fishery extension services that is mostly needed is group construction, while the extension performance that is considered the lowest is connecting fish processors to ousiders. According to the respondents, the highest extension performance is the extension worker's routine visits to their village (Table 5). When the suitability between the extension performance and the fish processors' importance is rated, it shows that the visit to other business sites is the lowest. This means that the extension method like field visit according to the respondents is very important; however, this has never been carried out. The field visit method is very important for the fish processors to get an experience from other fish processors and to increase their self-confidence to adopt innovation from other fish processors. As has been said that a farmer-to-farmer method is an effective learning process for farmers.

The respondents' satisfaction level on fishery extension in total shows a score of 67.5 which belongs to the category of "quite satisfied" (Table 5). The highest scored attributes on satisfaction are routine extension visits, fish processor group construction, the involvement of extension workers in solving business problems.

3.1 *The extension strategies of needs-based fishery processing using matrix model*

Needs-based fishery extension strategies are made according to the results of matrix analysis of innovation performance and extension services. Each quadrant shows one strategy. Quadrant I shows the main strategy (concentrate here) where the atrribute is in the condition of the highest importance and low performance; Quadrant II (keep up the good work) is in the position of high importance and high performance; Quadrant III (low priority), low importance and low performance; and Quadrant IV (possible overkill), low importance and high performance.

Table 4. The importance, performance, and satisfaction scores of the fish processors on fish processing innovation in Cirebon.

No	Attribute	Average scores of importance (Y)	Weighting Importance Score	Rank	Average scores of achieve ment (X)	Weighting Satisfaction Score	Rank	Suitability rate	Rank
1	Availability of alternative raw materials	3.55	10.88	2	2.97	0.32	1	83.66	2
2	Availability of additional materials	3.46	10.61	3	2.82	0.30	3	81.50	3
3	New processing machines	3.02	9.26	8	2.31	0.21	8	76.49	6
4	New ways of efficient processing	3.35	10.27	6	2.68	0.28	4	80.00	4
5	Ways to get clean water	3.42	10.48	5	2.97	0.31	2	86.84	1
6	New ways to handle waste	3.43	10.52	4	2.57	0.27	5	74.93	7
7	New packaging tools/materials	3.31	10.15	7	2.38	0.24	7	71.90	9
8	Ways to get businesss capital	3.57	10.94	1	2.38	0.26	6	66.67	10
9	Use of internet as promotion media	2.78	8.52	9	2.08	0.18	9	74.82	8
10	Use of conventional promotion media	2.73	8.37	10	2.13	0.18	9	78.02	5
	Total average	32.62	100		25.29				
	Weighting Average Total					2.55			
	Customer Satisfaction Index*					63.75			

* score 0-25=unsatisfied, 26-50=less satisfied, 51-75=quite satisfied and 76-100=satisfied

Table 5. The importance, performance, and satisfaction scores of the fish processors on extension services in Cirebon.

No	Attribute	Average scores of importance (Y)	Weighting Importance Score	Rank	Average scores of achieve ment (X)	Weighting Satisfaction Score	Rank	Suitability rate	Rank
1	Extension worker involvement in getting capital	3.57	10.44	2	2,.6	0.27	7	71.71	9
2	Group training by extension worker	3.58	10.47	1	3.00	0.31	2	83.80	2
3	New and suitable extension materials	3.38	9.88	6	2.68	0.26	6	79.29	5
4	Extension worker's routine visits	3.47	1.15	4	3.08	0.31	1	88.76	1
5	Fish processing training	3.41	9.97	5	2.58	0.26	4	75.66	8
6	Visits to other business locatins	3.35	9.80	7	2.35	0.23	9	70.15	10
7	New information from the extension worker	3.30	9.65	9	2.71	0.26	5	82.12	4
8	Connecting with the extension worker's fellows	3.31	9.68	8	2.53	0.24	8	76.44	6
9	Connecting with other parties	3.26	9.53	10	2.47	0.24	10	75.77	7
10	Involvement in solving problems	3.57	10.44	3	2.96	0.31	3	82.91	3
	Average Total	34.20	100.00		26.92				
	Weighting Average Total					2.70			
	Customer Satisfaction Index*)					67.50			

* score 0-25=unsatisfied, 26-50=less satisfied, 51-75=quite satisfied and 76-100=satisfied

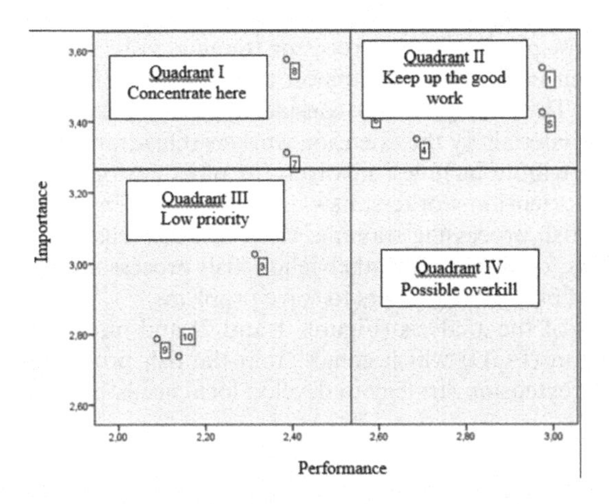

Figure 1. Strategy to increase the sustainability of fish processing business through processing innovation in Cirebon.

The main strategy that can be applied to reach the goal of traditional fish processing business sustainability can be seen in Figure 1. The priorities of the main strategy are the availability of business capital and packaging of processed fish products (quadrant I). The strategy that needs to be maintained is the availability of alternative raw materials, extra materials, new methods for efficient processing, methods for obtaining clean water, and new methods for handling waste (quadrant II). The strategy that has low priority is new processing machines, using internet media for promotion, and using conventional media for promotion.

Innovation attribute includes: alternative raw materials, extra materials, new processing machines, new methods for efficient processing, methods for obtaining clear water, new methods for handling waste, new packaging material/device, ways of getting business capital, using internet media for promotion, and using conventional media for promotion.

The strategy of extension services to increase the sustainability of fish processing business as can be seen in Figure 2 is a priority strategy that involves the extension workers to access capital for the fish processors (Quadrant I). This means that the fish processors need the extension workers to help them get access for their capital to a financial institution, namely very low interest. The strategy that needs to be maintained is developing the fish processors continuously and asking for the extension workers' advice to help fish procesors to solve problems (Quadrant II).

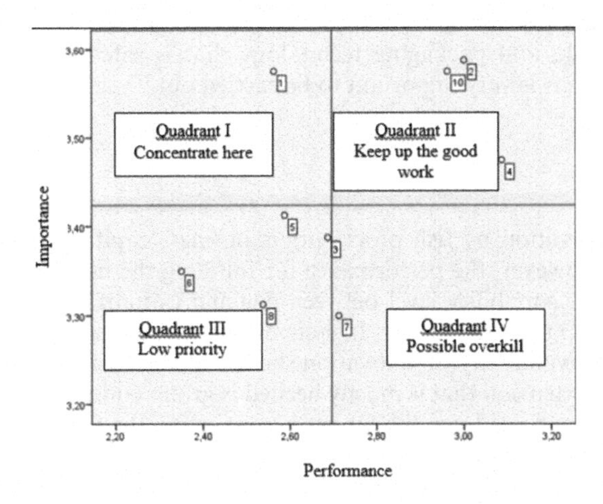

Figure 2. Strategy to increase the sustainability of fish processing business through extension services in Cirebon.

The strategy that has low priority is fish processor training, visits to other business's sites, and involvement of the extension workers to connect them to fellow fish processors and to stakeholders (Quadrant III). The strategy that is considered excessive is giving information and new and suitable extension materials by the extension workers (Quadrant IV).

Extension services attribute includes: involvement of extension workers to obtain capital, group construction by extension workers, new and suitable extension materials, routine visits by extension workers, fish processing training, visits to other business sites, new information from extension workers, connecting to other fellow fish processors, connecting to other parties, and involvement of extension workers to solve problems.

Based on the results of the analysis (Figures 1 and 2) and information from the result of Focus Group Discussion (FGD) which comes from the fish processors themselves, the formula of fish processing extension strategy to develop local needs-based business is as follows:

1. Increasing the role of extension workers

The role of an extension worker has proved to be important for the fish processors to develop their business. This role is not only transferring technology innovation to the fish processors, but more than that. The fish processors need the extension workers to facilitate them to get access to capital, to help them solve problems, to connect them to their fellow fish processors and stakeholders, and to develop group construction as a place for them to learn together.

2. Developing extension materials

The extension materials need to be developed, not only in terms of production techniques, but also in terms of post-production, especially qualified product packaging. With qualified packaging, the quality of the products can be maintained, and these products can reach wider markets. The materials for getting access to resources, especially capital, are needed because so far the fish processors are not well informed about the procedures to get capital loans from a financial institution.

3. Developing extension methods

Extension methods need to be developed, not only to change the participants' cognitive views, but also to improve their self-confidence, for example by visiting other fish processors' business sites outside their village, and they can share their experience with other fish processors as well.

4. Developing fish processing innovation

Innovation needed by traditional fish processors is not only technological innovation on production process, but also management innovation. Based on the rates of innovation needs, in fact the most needed innovation by fish processors is to get access to capital from a financial institution or bank, and packaging technology that is safe and interesting. Therefore, extension on both aspects is very important to be carried out.

4 CONCLUSIONS

The most needed innovation by fish processors is business capital with low interest from a financial institution; however, the performance for fulfilling the innovative capital is still low. Therefore, the score of suitability level between demand (importance) and supply (performance) for innovative capital is the lowest. In spite of that, in general the satisfaction rate of the fish processors on the availability of innovation shows "quite satisfied".

The type of fishery extension that is mostly needed is group construction, whereas the extension performance that is considered the lowest is connecting the fish processors to other parties outside the village. When rated from the suitability aspect between the extension performance and the needs of the fish processors, it can be seen that visits to other business sites attribute has the lowest rate. The satisfaction level of the respondents on extension services in general shows a "quite satisfied" category.

The main strategy to reach the goal of the sustainability of traditional fish processor business in terms of innovation is the availability of business capital with low interest and qualified packaging of processed fish products. In terms of extension services, the prioritized strategy is to increase the involvement of extension workers in getting an access to business capital. In general the strategies that need to be carried out are increasing the role of the extension workers, developing extension materials and extension methods, and developing fishery innovations.

REFERENCES

Devi, K.P.T., Suamba, I.K., & Artini, N.W.P. 2016. Analisis pengendalian mutu pada pengolahan ikan palagis beku. *Jurnal Agribisnis dan Agrowisata* 5(1):1-11

Esterlina, E.E. & Nanlohy, M. 2014. Analysis Of Total Bacteria in The Tuna Fish Asap Soaked with Liquid Smoke "Waa Sago" during Storage at Room Temperature. *Majalah Biam* 10 (2): 90-95.

FAO. 1992. Fish Processing Characteristics. Fermented fish in Africa: A study on processing marketing and consumption. http://www.fao.org/docrep/t0685e/T0685E05.htm

Fishery and Marine Office Cirebon Regency. 2015. Data Sentra Pengembangan Usaha Budidaya Air Tawar.

Holma, K., Ayinsa, & Maalekuu, B.K. 2013. Effect of traditional fish processing methods on the proximate composition of red fish stored under ambient room conditions. *American Journal of Food And Nutrition Print* 3 (2): 73-82.

Hudaya, Y., Hubeis, A.V., Sugihen. B.G., Fatchiya, A. 2017. The Effect of Extension on Small Scale Fish Processing Creativity in West Java, Indonesia. *Journal of Education and Practice* 8 (30): 150-153.

Junianingsih I. 2014. Strategi pengembangan usaha pengolahan tradisional ikan tongkol asap di Desa Jangkar Kabupaten Situbondo. *Samakia: Jurnal Ilmu Perikanan* 5(1): 31-38.

Kumolu-Johnson, C. A., Aladetohun, N. F., & Ndimele, P. E. 2010. The effects of smoking on the nutritional qualities and shelf-life of Clarias gariepinus. *African Journal of Biotechnology* 9 (1): 073-076.

Leeuwis, C. 2009. *Komunikasi untuk Inovasi Pedesaan. Berpikir Kembali tentang Penyuluhan Pedesaan.* Translation of Communication for Rural Innovation, Rethinking Agricultural Extension. Yogyakarta: Kanisius.

Lumban, R.M., Maulina, I., Gumilar, I. 2012. Analisis pengembangan usaha pemindangan ikan di Kecamatan Bekasi Barat. *Jurnal Perikanan dan Kelautan* 3(1): 17-24

Ministry of Marine and Fisheries of Indonesia. 2014. Sebaran UPI Skala Usaha Mikro Kecil Menengah (UMKM). http://www.djpdspkp.kkp.go.id/editor/gambar/file/PETA%20VOLUME%20PRODUK%20(01%20Desember%202014)%20baru.pdf

Njai, S.E. 2000. *Traditional Fish Processing And Marketing Of The Gambia.* Final Project 2000. Iceland: The United Nation University. http://www.unuftp.is/static/fellows/document/sirra3.pdf

Nti, C. A., Plahar W. A., & Patience Mateko Larweh. 2002. Impact of adoption in Ghana of an improved fish processing technology on household income, health and nutrition. *International Journal of Consumer Studies* 26:102–108.

Salán, O.E., Juliana, A.G., Marilia, O. 2006. Use of Smoking to Add Value to Salmoned Trout. *Brazilian Archives of Biology and Technology* 49(1): 57-62.

Sever, I. 2015. Importance-Performance Analysis: A Valid Management Tools? *Tourism Management* 48:43-53.

Supranto. 2011. *Pengukuran Tingkat Kepuasan Pelanggan Untuk Menaikkan Pangsa Pasar.* 4th ed. Jakarta: PT Rineka Cipta.

Suwardane, K.I., Fatchiya, A., & Sugihen, B.S. 2018. Capacity Building Of Ready To Serve Fish Processor In Micro Enterprises For Business Continuity In Pringsewu District. *International Journal of Social Science and Economic Research* (IJSSER) 3 (8):4267-4286.

Rogers, E.M. 2003. *Diffusion of Innovations.* Fifth Ed. New York: The Free Press.

Tutuarima, T. 2016. Total Plate Count on Soked Catfish in Panorama Market Bengkulu City During Storage Room Temperature. *Jurnal Agroindustri* 6 (1): 28-33.

Yanfika, H., Amanah, S., Fatchiya, A., & Asngari, P.S. 2018. Worker Performance From Perspective of Profit, Quality and Work Accuracy in Traditional Fishery Business in Lampung Province. *International Journal of Social Science and Economic Research* 3 (2):578-589.

Van den Ban, A.W. & Hawkisn, H.S. 1996. Agricultural Extension. California: Blackwell Science.

Rural Socio-Economic Transformation – Kinseng et al. (Eds)
© 2019 Taylor & Francis Group, London, ISBN 978-0-367-23603-8

Breastfeeding and online social support in a Facebook group

W. Yasya
University of Muhammadiyah Jakarta, Banten, Indonesia

P. Muljono, K.B. Seminar & Hardinsyah
Bogor Agricultural University, Bogor, West Java, Indonesia

ABSTRACT: Breastfeeding is known to positively affect maternal and child health, hence supporting the fulfilment of Sustainable Development Goals. Social support is a positive factor in increasing breastfeeding coverage. As information and communication technology advances, social support has extended online through social media such as Facebook. The purpose of this study was to analyze the relationship between online social support and adherence to breastfeeding behavior. The study used a quantitative approach with purposive sampling method on Facebook group members of the Indonesian Breastfeeding Mothers Association. The results showed significant relationship between online social support and breastfeeding behavior, with communication activeness and communication effectiveness having positive correlations while support access pattern, social support level and types of support communicated had no correlations with breastfeeding behavior. Thus it can be concluded that the higher social support obtained online through Facebook, the higher the adherence of breastfeeding behavior.

1 INTRODUCTION

1.1 *Rationale*

Breastfeeding is seen as a viable strategy in improving maternal and child health and consequently improve health, social and economic development. Numerous studies have shown that breastfeeding offers many benefits both for the child and the mother (Victora et al. 2016, Hahn-Holbrook et al. 2013, Binns et al. 2016). Victora et al. (2016) further asserts increasing breastfeeding worldwide can reduce 823,000 child deaths and 20,000 breast cancer deaths each year. Breastfeeding not only offers health benefits but may also have positive impact on the economy. If Indonesians engage in optimal breastfeeding practices, the country can conserve families up to 13.7% of monthly earnings that was to be spent on buying formula and save further to treat diarrhea and pneumonia, and in numbers it could save 256,420,000 USD annually in health system expenditures and prevent 1,343,700,000 USD in annual wage losses by improving children's cognitive abilities (Walters et al. 2016). In the long run, investing in breastfeeding will result in healthier, stronger and more productive adults for generations to come (Victora et al. 2016). Furthermore, WHO and UNICEF have identified that the scaling-up of breastfeeding may support the fulfilment of Sustainable Development Goals, namely goals 1, 2, 3, 4, 5, 8 and 10, resulting in a "healthier, more equitable — and thus, more sustainable – future" (Chan & Lake 2016).

Breastfeeding has been practiced since the dawn of mankind as a natural phenomenon to ensure the survival of their generation. For thousands of years, breast milk is a human baby's main food that is obtained directly from the mother's breast (Riordan & Wambach 2010). Breastfeeding is not a lifestyle choice but a normal part of human life (Calvert 2014). Breastfeeding is so important that in Islam—the major religion practiced by Indonesians—it is regarded as an obligation for mothers to breastfeed for two years and for fathers to ensure the child receives their right to breastfeed (Qur'an 2: 233). Recent research has led the World Health Organization to recommend exclusive breastfeeding – that is when the infant consumes only breast milk for the first six months of life – as one of the gold standard of infant and young child feeding, amongst early initiation of breastfeeding, adequate complementary feeding after six months and continuing breastfeeding for two years (World Health Organization 2003a). In Indonesia, the government has made exclusive breastfeeding mandatory unless there are medical reasons not to do so; those who intentionally hinder breastfeeding may face imprisonment for up to 1 year and be fined to a maximum of USD 10,000 according to the Republic of Indonesia Health Bill (Soekarjo & Zehner 2011). Specifically, the government has issued Government Decree on Exclusive Breastfeeding with several ministerial decrees to ensure the community supports exclusive breastfeeding by providing facilities at workplaces, health centers or any public setting.

Despite the many reported benefits and the obligation from religion as well as through government policy, while 49.9% of Indonesian babies were ever breastfed, only 37.3% were exclusively breastfed (Kementerian Kesehatan 2018). The number is still below the global target which is 50% exclusive breastfeeding (World Health Organization 2015b). Interactions between demographic, biological, psychological, and social factors contribute to the inadequate rate of breastfeeding (Kurniawan 2013). While demographic and biological problems are difficult to overcome, psychosocial factors of breastfeeding can be modified through behavior change communication, specifically health communication. In the age of networked communication technologies, health communication has evolved to incorporate new media as a main component to reach their communication goals.

Social media, the most recent form of new media, is seen to be beneficial in communicating health issues targeting women and specifically mothers. Women's communities have moved to the virtual space thus exchange of support and the need to connect are now facilitated by technology such as social media (Valtchanov et al. 2014). Community health communication are increasingly mediated by technology, replacing geographically-bound traditional tight-knit communities that seemed to fade due to higher mobility of women who are working and engaging more activity outside the home (Drentea & Moren-Cross 2005). Social media is readily available at any time, which is useful for mothers who feel isolated at home due to child-rearing to gain support, connections, advice and social interaction with other adults (Valtchanov et al. 2014, Gibson & Hanson 2013). The experience of motherhood mediated by Internet and social media allows women to feel connected with information and support from their virtual networks (Valtchanov et al. 2014). Research on technology use for breastfeeding communication revealed that technology can impact breastfeeding behavior through the provision of information and support. Information and support received online can help mothers with infants to breastfeed exclusively and minimize the use of formula (Newby et al. 2015). In the case of Indonesia, social media, specifically Facebook, is a potential tool for health communication since social media use accounts for 87.13% of Internet use even though the national Internet penetration still stands at 54.68% (APJII 2018). Furthermore, Indonesia is the fourth largest country on Facebook, and Facebook use still surpasses all other social media (Statista 2018).

From the above information we can see that health communication can influence attitudes, perception, knowledge, and social norms which act as precursors in behavior change. Health communication is the study of communication principles, processes and messages to provide health solutions at different levels (Dutta 2009). It focuses on how human and mediated communication impacts health care delivery and health promotion (Kreps et al. 1998). Communication then becomes a key component in behavior change as communication involves creating, collecting and sharing health information. This communication process also allows persuasive messages to be devised and then disseminated through major channels to provide the target audience the relevant health information that can positively influence their health knowledge, attitude and behavior (Neuhauser & Kreps 2003). Therefore, studying communication behavior is a crucial part in a health behavior change intervention.

Social support is a communication behavior, where it is perceived that social support is the verbal and nonverbal communication between sender and receiver which reduces uncertainty about a situation, self, other self, or relationships, and functions to increase perception on personal control of someone's experience (Albrecht & Alderman 1987, Albrecht & Goldsmith, 2003). Extensive literature have reported that social support has positive influence on health (Braithwaite et al. 1999, Gray 2013, Oh et al. 2013). Social support is broadly defined as the exchange of resource or assistance by members in a community (Cohen & Hoberman 1983). It can be categorized into two broad types: "nurturant" support and "action facilitating" support (Cutrona & Suhr 1992). Nurturant support is the type that helps with coping without necessarily solving the problem hence more emotional in nature, giving encouragement and increasing self-esteem, whereas action facilitating support intends to solve the problem causing stress through giving information or tangible actions (Selby et al. 2010). Social support is vital in determining breastfeeding success. In the era where many mothers are digital natives then the manifestation of social support has moved to cyberspace facilitated by technology (Gray 2013, Audelo 2014, Valtchanov et al. 2014). Thus the term "online social support" was coined to refer to the social support uniquely obtained through the Internet. Burman (2013) found that online social support can be an alternative to increase breastfeeding self-efficacy – another psychosocial factor vital in breastfeeding. Through a mixed-methods content analysis on an online breastfeeding forum, Gray (2013) identified the types of social support and the information sought or given through those support. It was found that social support in a breastfeeding forum were informational in nature and centered towards managing physical breastfeeding struggles (Gray 2013).

1.2 *Research question and hypothesis*

Based on the rationale and previous research on the subject as described in the previous section, the study was conducted with the aim to investigate the relationship between online social support and breastfeeding behavior in a Facebook group. In doing so, the following research question served as a guide in the research what is the relationship between online social support and breastfeeding behavior? In answering the research question, we proposed a hypothesis to be tested which is as follows:

> *H_1. There is a significant correlation between online social support and breastfeeding behavior.*

2 METHOD

2.1 *Methodology and sampling*

The study used quantitative methodology with positivistic approach to address the research questions. Research was conducted on April to June 2018. Data collection was done using online survey. Samples were selected using purposive sampling method on the members of Indonesian Breastfeeding Mothers Association (*Asosiasi Ibu Menyusui Indonesia* AIMI) which is the largest groups on breastfeeding in Bahasa Indonesia with about 219,000 members at the time of research. Sample size was obtained by taking into account the data needed for a correlational analysis with is a minimum of 10 times the number of variables to be computed or a minimum of 30 (Siddiqui 2013, Sugiyono 2013). The final sample consisted of 66 respondents selected to be analyzed.

2.2 *Instruments and data analysis*

The study used a questionnaire with open and closed questions on: 1) online social support with indicators measuring support access patterns, communication activeness, communication effectiveness and the type of support communicated; 2) breastfeeding behavior which consist of attitudes towards breastfeeding, breastfeeding self-efficacy and breastfeeding practice. The online social support questionnaire was developed by the researchers. Breastfeeding behavior measurement used questionnaires developed by researchers for measuring breastfeeding knowledge based on the basic concepts of lactation (Cadwell et al. 2006) and breastfeeding practice questions were based on the adherence in following the WHO and UNICEF Golden Standard on Infant and Young Child Feeding (2003a). Questionnaires for breastfeeding behavior was adapted from previously tested instruments which are the Iowa Infant Feeding Attitude Scale (IIFAS) by de la Mora et al.(1999) to measure attitude toward breastfeeding and the Breastfeeding Self-Efficacy Scale-Short Form (BSES-SF) developed by Dennis (2003) to measure breastfeeding self-efficacy. All instruments were translated into Bahasa Indonesia and the questions were worded to suit Indonesian readers. The instruments' validity (Spearman's r) range from 0.255 – 0.931 > $r_{0.05}$ = 0.374 and the reliability (alpha coefficient) is 0.865 for the online social support variable and 0.824 for the breastfeeding behavior variable, showing the instrument is valid and reliable to be used. The data obtained from the questionnaires were analyzed using SPSS 23 for descriptive statistics and correlational analysis using Spearman's Rank method.

3 RESULTS AND DISCUSSION

3.1 *Results*

3.1.1 *Respondent characteristics*

The respondents in this research were mothers aged between 20 to 29 years old (N=30, 45.5%), 30 to 39 years old (N=35, 53%) and 1 aged over 40 years old (1.5%) with high educational qualifications that are Diploma/Bachelors (N=49, 74.2%) and Postgraduate (N=10, 15.2%). Most of the respondents had infant children aged 0-6 months (N=31, 47.0%) while others had children aged 7-12 months (N=12, 18.2%), 13-24 months (N=17, 25.8%) and over 24 months (N=6, 9.1%). The respondents were mainly working mothers with full time employment (N=32, 48.5%) and part time employment (N=9, 13.6%), while the rest were unemployed (N=24, 37.9%). The majority of respondents were new to breastfeeding (N=37, 56.1%) while 43.9% (N=29) had previous breastfeeding experience.

3.1.2 *Online social support*

Online social support was measured using 5 indicators namely: 1) support access pattern, 2) communication activeness, 3) communication effectiveness, 4) social support level and 5) type of support communicated. Support access pattern measured how long the respondent had been a member of the Facebook group, how they accessed the group (directly or indirectly through News Feed) and how often they accessed the group in a week. Results showed that 63% were in the medium range of support access pattern. Communication activeness measured how often they and how often interacted in exchanging support through publishing posts, comments, or Reactions (Likes), which gave the majority reported having low communication activeness (36%). However the majority of respondents reported the communication effectiveness as high (56%) and the social support level also as high (46%). The type of support communicated in the group were perceived as more nurturing (56%) rather than action-facilitating support (44%).

3.1.3 *Breastfeeding behavior*

Breastfeeding behavior was measured using 4 indicators: 1) breastfeeding knowledge, 2) attitude towards breastfeeding, 3) breastfeeding self-efficacy, and 4) breastfeeding practice. The respondents had high level of knowledge (94%), were mostly positive about their attitude towards breastfeeding (67%) and reported high self-efficacy (61%).

The majority of respondents had favorable breastfeeding practice as reported by the high number of respondents who practiced exclusive breastfeeding (88%) and beginning solid feeding only after their infant is 6 months old (100%). However, only 26% reported practicing early initiation of breastfeeding during birth and only 29% reported practicing breastfeeding for 2 years, although the latter result were probably due to the fact that the majority of respondents were still breastfeeding infants less than 12 months hence this question did not apply to them (66%).

3.1.4 *Association of online social support and breastfeeding behavior*
Results from correlational analysis using Spearman's Rank method showed the correlation coefficient r=0.251 with p=0.042 < 0.05 thus the hypothesis is accepted. We further examined the association between the indicators of online social support to see which ones had correlations with breastfeeding behavior. The results of the correlational analysis is presented in Table 1. Based on the results, the indicators of online social support which had associations with breastfeeding behavior were communication activeness and communication effectiveness. While support access pattern, social support level and type of support communicated did not have any correlation to breastfeeding behavior.

3.2 *Discussion*

Results showed that there is a weak but positive correlation between online social support with breastfeeding behavior. This proves Bandura's theory on health promotion using social cognitive theory (Bandura 1998). According to Bandura, using social cognitive theory, there are two ways in which health communication can shape behavior: the direct pathway and connected through social systems. In the direct pathway, communication media fosters change by informing, modeling, motivating and guiding personal changes, whereas in the socially mediated pathway, communication media connected people to social networks and communities which provided personalized guidance, natural incentives and social support for the desired behavior change (Bandura 2004). According to Bandura, most changes occurred in the socially mediated pathway (Bandura 2009). The interesting phenomena about social media is that it not only served the purpose of informing, modeling, motivating and guiding as in the direct pathway, but the interactive nature of social media makes it also influencing behavior change through the socially mediated pathway, hence the positive association between online social support—that is uniquely obtained only through social media—and breastfeeding behavior.

Furthermore, the correlations between online social support and breastfeeding behavior were through communication activeness and communication effectiveness, whereas support access pattern, social support level and type of support communicated had no significant correlation with breastfeeding behavior. This could be an indication that the relationship between online social support and breastfeeding behavior were more strengthened by the quality rather than the quantity of the communication, since support access pattern that measured duration, frequency, and intensity and social support level were not correlated with breastfeeding behavior. What this means is that the level of engagement and the quality of interaction with each other mattered more in increasing adherence to breastfeeding behavior rather than merely accessing the site but not actively participating. Being active and engaged in social support communication can increase self-efficacy, thus increasing the positive behavior change (Bandura 2004). As proven by Oh et al. (2013), there are positive associations between having

Table 1. Correlation coefficient (r) of online social support and its indicators with breastfeeding behavior.

Indicators of online social support	Breastfeeding behavior
Online social support	0.251* (p=0.042)
– Support access pattern	0.108 (p=0.389)
– Communication activeness	0.349** (p=0.004)
– Communication effectiveness	0.248* (p=0.045)
– Social support level	0.224 (p=0.070)
– Type of support communicated	0.059 (p=-.376)

** statistically significant at 99%
* statistically significant at 95%

health concerns, seeking and perceiving social support from Facebook, and enhanced health self-efficacy, thus increasing their confidence in managing health, and in this case, in managing the success of breastfeeding.

High communication activeness level, marked by the level of interaction in the group such as how often they post, comment or give Reaction (Likes) and the intention of visiting the Group Site or click and read through personal News Feed (rather than just read through the News Feed then scroll down to read something else), was positively associated with breastfeeding behavior. This means that if the member is more active communicating and interacting in the group, then her breastfeeding behavior is also likely to be better. This is in line with the findings of Nurfirdauzi & Sutopo (2014) who studied the same Facebook group and found that the higher the role of AIMI Facebook group as communication medium, the higher the positive perception of breastfeeding mothers in implementing the exclusive breastfeeding program, in this case the perception is one that corresponds with scientific evidence from the government, researchers, and health professionals. Other than that, communication effectiveness which measured how far the level of support sought matches the level of support given, is also positively associated with breastfeeding behavior. The more effective the communication, meaning that the support given matches with the support sought, then the better the breastfeeding behavior. This proves Cutrona and Russell's *Optimal Matching Theory* where social support will be effective if the support given matches the support sought (Cutrona & Russell 1990). The results also indicated that the more effective the communication between members in AIMI group, the more likely their breastfeeding behavior improves.

Another important finding to note is that the respondents perceived the type of support communicated as more nurturing than action facilitating. The findings addresses the theoretical underpinnings of women's communication, where we have established that the nature of women's communication as generally supportive. Pew Research Center's report on parents' social media use found that mothers are more active on Facebook for the purpose of exchanging support on social media (Duggan et al. 2015). As suggested by Deetjen & Powell (2016), in online support groups for health where the participants were mostly women, emotional support became the most abundant type exchanged. This is again due to the nature of women's communication which are emotion-focused (Mo et al. 2009). Referring to the breastfeeding behavior results, the knowledge level of the respondents were reported as very high (94%), whereas attitude and self-efficacy were significantly lower at 67% and 61% respectively. This indicated that they already understand the basic concept of lactation and breastfeeding but just needed to interact and share support on Facebook to increase their attitude and self-efficacy, which in turn would improve their breastfeeding behavior.

4 CONCLUSION

Based on the analysis presented in the previous sections, it can be concluded that online social support has positive significant association with breastfeeding behavior. The indicators of online social support namely communication activeness and communication effectiveness are significantly correlated with breastfeeding behavior, whereas support access pattern, social support level and types of support communicated had no correlation with breastfeeding behavior. Thus, we can conclude that Facebook use especially activeness in Facebook groups on breastfeeding can positively improve breastfeeding behavior, and that the quality of communication plays more part in determining the adherence of breastfeeding behavior compared to the quantity of communication.

It is expected that the results of this research can contribute in maximizing the use of social media in increasing the adherence to breastfeeding behavior, since little research had been reported that quantitatively associate social media use with breastfeeding. It is suggested that further research utilized more samples so that more variables that may affect breastfeeding can be used, such as individual characteristics, environmental/offline social support, and the patterns of general Facebook use.

REFERENCES

Albrecht, T.L. & Alderman, M.B. 1987. *Communicating social support*. Thousand Oaks: Sage.

Albrecht, T.L. & Goldsmith, D.J. 2003. Social Support, Social Networks and Health. In T. L. Thompson, A. Dorsey, R. Parrott, & K. Miller, eds. *Handbook of Health Communication*. Mahwah, NJ: Lawrence Erlbaum.

APJII. 2018. *Penetrasi dan Perilaku Pengguna Internet Indonesia*. Jakarta: Asosiasi Penyelenggara Jasa Internet Indonesia.

Bandura, A. 1998. Health promotion from the perspective of social cognitive theory. *Psychology & Health* 13(4): 623–649.

Bandura, A. 2004. Health promotion by social cognitive means. *Health Education & Behavior* 31(2): 143–164.

Bandura, A. 2009. Social Cognitive Theory of Mass Communication. In J. Bryant & M. Oliver, eds. *Media Effects: Advances in Theory and Research*. New York: Routledge.

Binns, C., Lee, M. & Low, W.Y. 2016. The Long-Term Public Health Benefits of Breastfeeding. *Asia-Pacific Journal of Public Health* 28(1): 7–14.

Braithwaite, D.O., Waldron, V.R. & Finn, J. 1999. Communication of social support in computer-mediated groups for people with disabilities. *Health communication* 11(2): 123–151.

Burman, A.B. 2013. Can Internet Applications Help with Feeding Babies? A Study of Online Social Support Amongst Breastfeeding Women. *Medicine 2.0 Conference*. http://www.medicine20congress.com/ocs/index.php/med/med2013/paper/view/1576 3 March 2016.

Cadwell, K., Turner-Maffei, C. & O'Connor, B. 2006. *Maternal and infant assessment for breastfeeding and human lactation: A guide for the practitioner*. Sudbury: Jones & Bartlett Learning.

Calvert, H. 2014. 'Breast isn't best, it's just normal'. *Nursing Children and Young People* 26(10): 15–15.

Chan, M. & Lake, A. 2016. Breastfeeding: A Key to Sustainable Development. *WHO and UNICEF World Breastfeeding Week 2016 Message*. http://who.int/mediacentre/events/2016/2016-world-breast feeding-week-letter.pdf?ua=1 29 October 2018.

Cohen, S. & Hoberman, H. 1983. Positive events and social supports as buffers of life change stress. *Journal of Applied Social Psychology* 13(2): 99–125.

Cutrona, C.E. & Russell, D.W. 1990. Type of social support and specific stress: Toward a theory of optimal matching. In B. R. Sarason, I. G. Sarason, & G. R. Pierce, eds. *Social support: An interactional view*: 319–366. Oxford: John Wiley & Sons.

Cutrona, C.E. & Suhr, J.A. 1992. Controllability of stressful events and satisfaction with spouse support behaviors. *Communication Research* 19(2): 154–174.

Deetjen, U. & Powell, J.A. 2016. Informational and emotional elements in online support groups: A Bayesian approach to large-scale content analysis. *Journal of the American Medical Informatics Association* 23(3): 508–513.

Dennis, C.-L. 2003. The Breastfeeding Self-Efficacy Scale: Psychometric Assessment of the Short Form. *Journal of Obstetric, Gynecologic, & Neonatal Nursing* 32(6): 734–744.

Drentea, P. & Moren-Cross, J.L. 2005. Social capital and social support on the web: The case of an internet mother site. *Sociology of Health and Illness* 27(7): 920–943.

Duggan, M., Lenhart, A., Lampe, C. & Ellison, N.B. 2015. *Parents and social media: mothers are especially likely to give and receive support on social media*.

Dutta, M.J. 2009. Health Communication: Trends and Future Directions. In J. Parker & E. Thorson, eds. *Health Communication in the New Media Landscape*: 59–92. New York: Springer.

Gibson, L. & Hanson, V.L. 2013. Digital motherhood. *Proceedings of the SIGCHI Conference on Human Factors in Computing* Systems - *CHI '13*: 313. http://dl.acm.org/citation.cfm?id=2470654.2470700.

Gray, J. 2013. Feeding On the Web: Online Social Support in the Breastfeeding Context. *Communication Research Reports* 30 (April 2015): 1–11.

Hahn-Holbrook, J., Schetter, C.D. & Haselton, M.G. 2013. The advantages and disadvantages of breastfeeding for maternal mental and physical health. In M. Spiers, G. P, & K. J, eds. *Women's Health Psychology*: 414–439. New Jersey: Wiley.

Kementerian Kesehatan. 2018. *Hasil Utama RISKESDAS 2018*. Jakarta: Balitbangkes Kementerian Kesehatan.

Kreps, G.L., Bonaguro, E.W. & Query, J.L. 1998. The History and Development of the Field of Health Communicatio. In L. D. Jackson & B. K. Duffy, eds. *Health Communication Research: Guide to Developments and Directions*: 1–7. Westport, CT: Greenwood Press.

Kurniawan, B. 2013. Determinan Keberhasilan Pemberian Air Susu Ibu Eksklusif. *Jurnal Kedokteran Brawijaya* 27(4): 236–240.

de la Mora, A., Russell, D., Dungy, C., Losch, M. & Dusdieker, L. 1999. The Iowa Infant Feeding Attitude Scale: Analysis of reliability and validity. *Journal of Applied Social Psychology*. 29: 2362–2380.
Mo, P.K.H., Malik, S.H. & Coulson, N.S. 2009. Gender differences in computer-mediated communication: A systematic literature review of online health-related support groups. *Patient Education and Counseling* 75(1): 16–24.
Neuhauser, L. & Kreps, G.L. 2003. Rethinking communication in the e-health era. *Journal of Health Psychology* 8(1): 7–23.
Newby, R., Brodribb, W., Ware, R.S. & Davies, P.S.W. 2015. Internet Use by First-Time Mothers for Infant Feeding Support. *Journal of human lactation : official journal of International Lactation Consultant Association* 31(3): 416–424.
Nurfirdauzi, R.A. & Sutopo. 2014. Peran Media Komunikasi Facebook Asosiasi Ibu Menyusui Indonesia Terhadap Persepsi Ibu Menyusui Dalam Melaksanakan Program ASI Eksklusif. *Journal of Rural and Development* 5:(215–225).
Oh, H.J., Lauckner, C., Boehmer, J., Fewins-Bliss, R. & Li, K. 2013. Facebooking for health: An examination into the solicitation and effects of health-related social support on social networking sites. *Computers in Human Behavior* 29(5): 2072–2080.
Riordan, J. & Wambach, K. 2010. *Breastfeeding and Human Lactation*. Sudbury: Jones & Bartlett Publishers.
Selby, P., Van Mierlo, T., Voci, S.C., Parent, D. & Cunningham, J.A. 2010. Online social and professional support for smokers trying to quit: An exploration of first time posts from 2562 members. *Journal of Medical Internet Research* 12(3): 1–12.
Siddiqui, K. 2013. Heuristics for sample size determination in multivariate statistical techniques. *World Applied Sciences Journal* 27(2): 285–287.
Soekarjo, D. & Zehner, E. 2011. Legislation should support optimal breastfeeding practices and access to low-cost, high-quality complementary foods: Indonesia provides a case study. *Maternal and Child Nutrition* 7(SUPPL. 3): 112–122.
Statista. 2018. Most popular social networks worldwide as of October 2018, ranked by number of active users (in millions). https://www.statista.com/statistics/272014/global-social-networks-ranked-by-number-of-users/10November2018.
Sugiyono. 2013. *Statistika untuk Penelitian*. Bandung: Alfabeta.
Valtchanov, B.L., Parry, D.C., Glover, T.D. & Mulcahy, C.M. 2014. Neighborhood at your Fingertips: Transforming Community Online through a Canadian Social Networking Site for Mothers. *Gender, Technology and Development* 18(2): 187–217.
Victora, C.G., Bahl, R., Barros, A.J.D., França, G.V.A., Horton, S., Krasevec, J., Murch, S., Sankar, M.J., Walker, N. & Rollins, N.C. 2016. Breastfeeding in the 21st century: epidemiology, mechanisms, and lifelong effect. *The Lancet* 387(10017): 475–490.
Walters, D., Horton, S., Siregar, A.Y.M., Pitriyan, P., Hajeebhoy, N., Mathisen, R., Phan, L.T.H. & Rudert, C. 2016. The cost of not breastfeeding in Southeast Asia. *Health Policy and Planning* 31(8): 1107–1116.
World Health Organization. 2003. *Global strategy for infant and young child feeding*. Geneva (CH): World Health Organization.
World Health Organization. 2015. WHO | Global targets 2025. *WHO*. http://www.who.int/nutrition/topics/nutrition_globaltargets2025/en/# 13 June 2016.

Rural Socio-Economic Transformation – Kinseng et al. (Eds)
© 2019 Taylor & Francis Group, London, ISBN 978-0-367-23603-8

Communication of the organizational culture in Village-Owned Enterprises (BUMDes) for sustainable rural entrepreneurship management based on local wisdom

S. Kusuma
School of Communication, Atma Jaya Catholic University, Jakarta, Indonesia

A.V.S. Hubeis, S. Sarwoprasodjo & B. Ginting
Department of Communication and Community Development Sciences, Bogor Agricultural University, Bogor, West Java, Indonesia

ABSTRACT: BUMDes of Panggungharjo Village used local wisdom to manage environment management business, agribusiness, and rural tourism. The objective of this study is to analyze the meaning of organizational culture in BUMDes Panggungharjo Village for a sustainable rural entrepreneurship management based on local wisdom. This study employed qualitative method and applied Ethnography of Communication design. The techniques of collecting data employed observations, FGDs, individual interviews, and study of relevant documents. The findings showed that BUMDes have communication performance creating organizational culture to explore idea and thought of members, innovation, participative plan, and information access transparency.

1 INTRODUCTIONS

Rural entrepreneurship is a solution to lower unemployment rates and reduce the economic problem in rural areas as a means of improving the quality of individual, family, and community's life contributing to a sustainable rural development (Mayer et al. 2016, Andrew 2015, Toosi et al. 2015, Das 2014, Korsgaard & Müller 2015). The location existing in the rural area benefits the businessmen and makes them interested in providing prosperity and development in rural areas. Rural entrepreneurship should never ignore the potency of rural entrepreneurs, despite its very slow and low growth rate (Korsgaard & Müller 2015).

Village-Owned Enterprises (Indonesian: *Badan Usaha Milik Desa*, thereafter called BUMDes) is a rural enterprise established by village (under Village regulation) corresponding to the village's need and potency. The need and potency are discussed through Village Discussion between Village Council (Indonesian: *Badan Permusyawaratan Desa*, thereafter called BPD), Village Government, and community elements. BUMDes can run business in economic and/or public service sector, managing asset, service, and other business as much as possible for the sake of Villagers' prosperity (Law No.32 of 2004, PP No.71 of 2005, Law No.6 of 2014, Minister of Village's Regulation [Permendesa] No.4 of 2015). Within 14 years since the issuance of Law of 2004 about BUMDes, there were only 1,022 BUMDes in 2014; this figure increases dramatically to 14,686 BUMDes in 2016 and to 39,149 BUMDes in 2018. However, out of 74,958 villages in Indonesia, many of them do not have BUMDes. Ministry of Village, Disadvantaged Regions, and Transmigration's Evaluation of 2017 shows that there are approximately eight thousand active BUMDes and only four thousands of them are profitable because of limited human resource, management, and productivity. It means that although regulations and fund grant exist, not all BUMDes can successfully operate their business sustainably—some even went bankrupt.

The problems founded in managing BUMDes sustainably were related to its performance, leadership, and management. Generally, BUMDes established from the government's imposition failed to succeed. The community sees BUMDes as a mere governmental project that has weak legitimacy and adhesiveness (Eko 2014). The development of rural entrepreneurship found some problems such as inadequate managerial skill, inadequate management comprehension and work planning, organizational skill and performance and entrepreneurship skill of local rural manager (Taghibeygi et al. 2015, Munyanyiwa & Mutsau 2015, Kashani et al. 2015). While leadership and management do not instantaneously make BUMDes become solid and sustainable, yet poor leadership and management will speed up the BUMDes' into suspended animation (Indonesian: *mati suri*) (Eko 2014).

This research assumes that the survival of BUMDes is determined by organizational culture including emotional and psychological climate or atmospheres such as employees' work ethos, attitude, and productivity (West & Turner 2013). Organizational culture also involves all symbols (action, routine activity, conversation, and so forth) and meanings labeled by people to these symbols. Members of the organization create and maintain a mutually shared feeling regarding the reality of the organization, leading to a better understanding of an organization's values. Individuals discuss with each other in creating and maintaining the reality of the organization (West & Turner 2013).

Considering 181 articles about rural entrepreneurship published in Scopus-indexed journals, it can be seen that organizational characteristic, policy measure, and framework of institution and government have attracted much attention. However, there are still very few articles on sociology, social science, and territorial studies (Pato & Teixeira 2014). Enterprise contributes to social-economic development of rural community supported by the government's public policy (Deshwal 2015, Nguyen & Frederick 2014, Lassithiotaki & Roubakou 2014, Phungwayo & Mogashoa 2014).

The organization of BUMDes, viewed from its management, performance, and leadership, has not run maximally yet due to inadequate facilitation from Provincial and Regency Governments (Safitri et al. 2016). The problem encountered by BUMDes is the administrators' poor knowledge on the management of BUMDes so that the performance of BUMDes in developing business is not optimal (Agunggunanto et al. 2016). The inhibiting factor is related to the strategy of managing BUMDes asset, particularly including difficulty in developing new business, limited innovation in developing local products, inadequate marketing infrastructure, and limited fund and support from both Regency and Provincial Governments (Hayyuna et al. 2014). Improvement needed by BUMDes is the service quality and ability to manage the organization. Poor communication and socialization made distrust arise among the members of BUMDEs and hindered the organization of BUMDes. These problems lead to the need for transparency and accountability in managing BUMDes (Anggraeni 2016). This research employs Organizational Culture theory to observe, to record, and to comprehend the communicative behavior of BUMDes organization members. This research studies the culture constructed communicatively through BUMDes' practices in the organization, including emotional and psychological climate or atmospheres such as employees' work ethos, attitude, and productivity. This research will analyze and answer two questions: (1) how is communication practice in BUMDes organization in Panggungharjo Village in Managing Village Enterprise? and (2) what is the meaning of organizational culture in BUMDes Panggungharjo for sustainable rural entrepreneurship management based on local wisdom?

This research aims to analyze the meaning and the perception of culture achieved through interactions occurred between members and management of BUMDes in Panggungharjo Village. Organizational culture includes all symbols (action, routine activity, conversation, and so forth) and the meaning labeled by people to these symbols. Organizational culture theory views emotional and psychological climate or atmosphere involving the employees' work ethos, attitude, and productivity of BUMDes organization as a village enterprise. Pacanowsky & O'Donnell Trujillo (1982) stated that the members of the organization perform certain communication leading to the appearance of unique organizational culture. Performance is a metaphor representing a symbolic process of understanding human behavior in an organization. Theoreticians explain five cultural performances: ritual, passion, social, political, and enculturation. These performances can be undertaken by any member of the organization. Cultural meaning and understanding are achieved through interactions occurred between members and management of BUMDes organization as a rural enterprise, where individuals discuss with each other through active participation in creating and maintaining the reality of organization and in finding the values of BUMDes organization.

Gani (2014) viewed entrepreneurship as a behavior rather than personality characteristic. Entrepreneurship is a working practice building on concept and theory rather than on intuition. Therefore, entrepreneurship can be studied and mastered in a systematic and planned manner. He recommended three behavior elements to be owned in order to support the successful entrepreneurship practice: 1) purposeful innovation, 2) enterprise management, and 3) enterprise strategy.

This research also studies local wisdom (Ife 2002) belonging to BUMDes. Local wisdom (Ife 2002) is the values created, developed, and maintained in the local community and its survivability makes it the guideline for its community's life. Local wisdom includes various mechanisms and ways of behaving and acting manifested into the social order. Local wisdom consists of five dimensions: local knowledge, local value, local skill, local resource, and local decision-making mechanism.

2 METHODS

This research employed a qualitative method with a constructivist paradigm. Constructivist paradigm, according to Denzin & Lincoln (2017), views the complex real experience world from the perspective of people living within it. World's reality and specific meanings of the situation are constructed by social actors and become the research object of the qualitative method. It means that certain actors in certain place and at a certain time give meaning to various events and phenomena through a long complex social interaction process involving history, language, and action (Patton 2002).

Creswell (2009) explains that social constructivism confirms the assumption that individuals always attempt to understand the world where they live and work. This research will investigate the process of establishing and developing BUMDes sustainably. This research conducted in two levels: an individual case study focuses on studying individuals, programs, events, or process resulting from a certain concept and an organizational case, a case study to obtain information about the organization (Yin 2009).

The unit of analysis is related to the problem of determining what the "case" means in the research (Yin 2009). The unit of analysis and the informant of research consisted of the members of BUMDes Panggung Lestari place in Panggungharjo Village, Sewon Sub District, Bantul Regency, Daerah Istimewa Yogyakarta Province. The BUMDes has run an innovative enterprise and management with two business units: Environment Management and Village Tourism Management Services. Environment Management Service business unit includes Tamanu Oil, Used Coconut Oil, and Rubbish Bank. Meanwhile, Village Tourism Management service includes culinary, agribusiness, Swadesa (Self-Service Village) and PKL (street sellers), that cooperates with Village Government and other private partners.

Ethnography of Communication method was used to analyze the pattern of communication behavior and social interaction of individuals or members of BUMDes Panggung Lestari in performing certain communication to build a unique organizational culture. This study also used Organizational Culture theory by observing the cultural performance including ritual performance (personal ritual, assignment ritual, social ritual, organizational ritual), passion performance, social performance, political performance, and enculturation performance.

Ethnography of Communication, according to Littlejohn & Foss (2017), is a method of applying simple ethnography method to a group's communication pattern. Primary data was obtained from Focus Group Discussions (FGDs) with BUMDes, observation on the BUMDes administrator meeting, and interviews with BUMDes administrators. Secondary data was obtained from relevant documents such as news, books, journal articles, and popular article.

The informants were selected using purposive sampling for the identification and selection of information-rich cases for the most effective use of limited resources (Patton 2002). A total

Table 1. The technique of collecting data and type of informants.

Type of Method	Total informant	Type of Informant
Focus Group Discussion (FGD) with 20 informants	20	Pak Lurah (Village Head), BUMDes supervisor (Village Representative Council or BPD), BUMDes Director, Operational Manager of Environment Management Service, Operational Manager of Village Tourism Management Service
Interview	20	BUMDes Secretary, BUMDes Treasurer, Head of BUMDes' General Division, Head of BUMDes' Procurement and Research and Development Division, Financial Staff of BUMDes, CEO of PT Tamanu Oil, Head of Tamanu Oil Unit Business and Used Coconut Oil, Head of Village and Market Business Unit, Head of HRD and Reservation Division, Head of Financial Division, Head of Village Supermarket (*Swadesa*) Business Unit, General Division, medium and export craftsperson, Administrators of Family Welfare Program (PKK) Administrators, Director of Rubbish Bank Business Group, Administrators of Dewi Kunti & PT. Martani Cooperatives.

of 20 informants participated in this research; FGD was conducted once with informants from BUMDes administrators, village government and stakeholders; meanwhile, interviews were conducted 20 times (Table 1).

3 RESULTS AND DISCUSSION

3.1 *Communication practice in BUMDes organization in Panggungharjo Village in managing village enterprise*

The director of BUMDes is aware of his limitation to communicate with the members of BUMDes, so he opened communication spaces, either personally or institutionally, where the employees have spaces to express their ideas and thoughts. The communication between Director and Manager is conducted to discuss the improvement of business production capacity, packaging of products, and income, and also finding solutions to problems encountered by business unit. The business unit head communicates with the employees to discuss objectives, program, work targets, and also to make the employees getting closer to each other so any discomfort on attitude or relations between employees could be lessened to support the organization of BUMDes.

Communication is conducted to establish a good structural relation with the management and to establish "silaturahmi" (Indonesian, meaning good relationship) just like a Javanese philosophy "nguwongke uwong" (humanizing human beings). It is also intended to help employees in other division to deliver information about work schedule, event agenda, and work hours, and to facilitate the job coordination between different business units. Additionally, it aims to ensure that work targets performed by the administrators of BUMDes.

All of the practices above are in line with he organizational cultural concept where organizational practices involving symbols and meanings are constructed communicatively. The manager of BUMDes business unit employs a variety of communicating approaches by visiting the employees in each division during his leisure time to discuss how to provide good service to consumers. The manager feels that communication will be more effective through group discussion according to the employees' job than through convening all divisions at the same time. The manager sees that during resting time, the employees are spread in some places thereby the customers could feel uncomfortable. To deal with this problem, the manager takes initiative to provide a special rest area for employees so that they can interact with each other there.

Pacanowsky & Trujillo (1982) suggested that the members of the organization perform certain communication leading to the appearance of unique organizational culture. Performance is a metaphor representing a symbolic process of understanding human behavior in an organization. All communication performances occurring regularly and repeatedly are called as ritual performance. Ritual is conducted both in verbal and nonverbal manner, usually taking the interaction with others into account. This article emphasizes on social ritual and organizational ritual. Social ritual can also involve nonverbal behavior. Organizational ritual is an organizational activity conducted frequently such as leader meeting and division meeting. Ritual, according to Littlejohn & Foss (2017) is something repeated regularly. Social rituals are not related to the task (assignment) but are an important appearance in an organization. Organization ritual is the one involving all workgroups with several regularities. The analysis on social ritual performance in BUMDes Panggung Lestari explained internal communication media, communication topic, communication objective and achievement, communication language and dialect, communication style, actor age, and actor involved in communication can be seen in Table 2.

Each business unit of BUMDes Panggung Lestari has its own strategy depending on its manager and unit head. The meetings in a business unit are related to monitoring and evaluation on activities, for example, to find out whether or not the employees have worked according to their job description and to find out the employees' performance. The management receives all reports from the employees including the problem between employees and the work-related problems which become topic to be discussed in order to prevent them from being prolonged problems. The analysis of the ritual performance of BUMDes Panggung Lestari can be seen in Table 3 below.

3.2 *The meaning of organizational culture in BUMDes Panggungharjo for sustainable rural entrepreneurship management based on local wisdom*

Basic knowledge of entrepreneurship is innovation. For that reason, a new way of utilizing resource should be communicated to create prosperity for the villagers. BUMDes Panggung Lestari builds enterprise by supporting the business of people surrounding Panggungharjo Village who start new business or old enterprises or SMEs that have not developed yet. The duty of management is, among others, to involve and to improve their business quality by accommodating food, beverage, and their work craft product and to discuss while exploring their actual needs, whether or not it needs capital, and whether the capital is in the form of money or goods. Therefore, personal approaches should be taken to facilitate them. BUMDes documents and registers the name and residence of entrepreneurs and then visit the entrepreneurs' houses.

Purposive innovation, enterprise management, and enterprise strategy are entrepreneurship behavior concepts. Communication behavior performance in constructing the BUMDes' organizational culture through interaction occurring between employees and management contributes to the management of sustainable rural entrepreneurial behavior. The manager of Kampoeng Mataram business unit invites their relations to a banquet so that they can know their village business and eventually will be interested in recommending others to come there. The second entrepreneurial marketing strategy is to rent this place for a wedding event and the third one is to promote it through social media. Meanwhile, this social media use is very limited because BUMDes wants objective broadcasting, and social media will represent BUMDes subjectively. However, considering the result of observation, social media can be an effective channel to promote this business, but it is the guests visiting that promote it through their social media.

BUMDes Panggung Lestari has natural resource only. A life landscape constituting the utilization of social landscape related to the social condition of society such as RPS (Rubbish Processing House), Tamanu oil and used oil processing constituting a technological landscape because it uses technological creativity approach. Cultural landscape in Kampoeng Mataram included the application and the delivery of understanding on society life in Panggungharjo Village, related to the high appreciation for art and culture.

In Indonesia, the only business producing Tamanu oil (*Minyak Nyamplung*) is BUMDes Panggungharjo. However, because BUMDes is not a legal entity, it is impossible for it to

Table 2. The analysis of social ritual performance in BUMDes Panggung Lestari.

No	Category	Findings
1.	Internal communication media	Communication is performed through WhatsApp (WA) media followed by Director, Secretary, Treasurer, the management staff of BUMDes, Managers of Environment Management Service and Village Tourism Management Service, Heads of Tamanu Oil and *Used Coconut Oil* business unit, Kampoeng Mataraman stall, and Street seller Swadesa business unit. There are 3 WA groups to communicate and to coordinate formally with Supervisor, Advisor, and Director of BUMDes and non-formal communication of administrators at management and business unit level.
2.	Communication topics	The problems occurring in business units are shared in non-formal WA group first in order to be known by all administrators and to make a decision; thereafter, they are shared in the formal group to be known by Pak Lurah (Village Head) and BPD. Information about consumers' problem should be followed-up directly by the administrators in the field. The management discusses the employees' grievances and problems limited to management level and performs technical coordination concerning the work and meeting schedule
3.	Communication objective and achievement	Informing and announcing the activities held in the management office and business unit, reminding the employee to be present, coordinating the activities in BUMDes and Village Government and informing the people's grievances delivered by Pak Lurah. Reminding each other about job matter, job evaluation, giving motivation, encouraging, struggling together to prevent the employees from feeling saturated in BUMDes organization.
4.	Communication language and dialect	Using formal languages (Indonesian language and soft Javanese language) in the office and formal WA group with Pak Lurah. Meanwhile, in other WA group, the informal language "*Bahasa Kekancan*" (friendship language) can be used to facilitate communication. Sometimes Pak Lurah communicates either formally and non-formally by making a joke. Junior employees use polite formal language to communicate with the senior ones.
5.	Communication style	Egalitarian communication style is used, in which all employees give recommendation and input to each other when any employees encounter a problem. Open communication style is also used, by which everyone can propose an idea or thought to discuss probabilities from many aspects so that the idea can be implemented collectively, as the trigger to find early input, target, and to formulate the plan concept to be done. Additionally, a two-way communication style is employed, by which the manager emphasizes more on the assumption that everything cannot always result from the top manager, suggestion or critique can also be given in a bottom-up manner. If the employees consider that Director of BUMDes' policy is inappropriate, they can give suggestion and input, and the Director should be willing to accept it openly to correct the error or weakness.
6.	Actor Age Actors involved in communication	Starting from the business unit head aged 20-years to Director of BUMDes and business unit head aged 40-year and the mean age of employees is 20 – 30 years. All employees are involved in communication to make innovation for the sake of business sustainability. Director of BUMDes does not limit but explore innovation from the bottom to build employees' loyalty and contribution to the company, although not all employees participate because they have no cellular phone.
7.	Obstacles	WA group of Kampoeng Mataram business unit employees have existed even before the new Unit Head came, formerly the group atmosphere is relaxed and cheerful, however, when new Unit Head joining the WA group, the group became not relaxed and desolate. This lead to a new WA group made by employees excluding the new Unit Head. Not all employees have an Android cellular phone, and even those having android cellular phone found difficulty in buying quota so that not all employees and just those having communication set only participate in WA group

Table 3. The analysis of the organizational ritual performance of BUMDes Panggung Lestari.

No	Category	Findings
1.	Meeting activity	The meeting between manager and business unit heads is held every week and the meeting of BUMDes management is held once a month. Internal management meeting is held on Saturday for coordination and weekly report in the form of work achievement in one last week. On Monday the administrators submit the work plan report for the next week to the Director. The manager usually discussed strategic matters, program development, and external relationship in this meeting. Technical meetings are usually held by Unit Head, where small groups are included in this meetings and held by the coordinator because the field supervision should be conducted every day to be synchronized with the employees' and the guest's grievances. This meeting discusses collectively work report, target, achievement, work process, the obstacles with the realization of relationship with consumers, and other problems in the evaluation meeting.
2.	Communication topic	Routine meeting at BUMDES management level discusses the target to make potential income from January to December. Achieved and unachieved target are discussed. Director of BUMDes emphasizes that the employees should be able to achieve the target because it will impact on salary cutting to maintain the work rhythm, the safe point of potential income, for-profit or not-for-profit, and to find out whether or not the operation of each business unit can be made efficient. This meeting also discusses gain, profit, target, complaint, need, weakness, such as fleet necessity, and finds out whether or not the weekly job has been corresponding to the plan. Usually, the manager needs a budget proposal to discuss the business unit and management.
3.	Language and dialect	The language used is flexible, it can be Javanese, mixed Javanese and Indonesian, informal but polite.
4.	Communication style	Equal communication style is used where subordinate and superior discussed with mutual respect principle. Democratic communication style is also used, starting with the elaboration of management, a one-way explanation from the Director first, and then individual units present their business development and ask for response openly, in a bottom-up and participatory manner.
5.	Men and women participation	Male employees said that women speak up more than men in the management because when some problems occurred, the women are more likely to express it in the meeting. In the meetings of Kampoeng Mataram business unit, women are usually more active. Meanwhile, the women employees feel that the male ones give more suggestions and recommendations because their number is larger than the female employees' and during the reporting meeting, women usually speak a little. Manager perceives that all male and female employees can express their opinion and there is no problem or no difference. Everyone can participate and none is more dominant.
6.	How to involve the members	All employees get opportunities and are obliged to speak up corresponding to their own position. Director of BUMDes offers the employees to give input and to report the constraints they encounter, and women usually express more problems. The monthly meeting is the time the employees share problems. Business unit head emphasizes on and attempts to apply participatory/democratic communication.
7.	The social status of actors	The communication between the chairperson of BPD and the management runs not too rigidly and he can communicate smartly with the employees. Director of BUMDes says that the position in the organization affects significantly to the communication between superior and subordinate because sometimes the employees will be passive when there is no stimulation. Director emphasizes that the position is only a mandate (trusteeship) so that in communicating the employees are expected to respect each other according to their own role and job position structurally. Beyond the workplace, the director will maintain a close friendship just like relatives and friends. Communication between superior and subordinate is very close, they can make a joke to establish intimacy and compactness, and there is no distance in a family because they feel comfortable with each other.

(Continued)

Table 3. (*Continued*)

No	Category	Findings
8.	Obstacles	Coordination meeting is conducted rarely unless there is emergency because many guests come on Saturday and Director of BUMDes is doing many jobs outside. Work hour in Saturday is only a half of day and the day will only be used for coordination meeting or weekly report meeting and the meeting for discussing the work plan for the next week. The weekly and monthly report meetings have been held several times, however it is not effective since there are members who have to work outside the office. Organizational regulation should be written and socialized clearly to the employees.

access fund from the third party. Therefore, on the initiative of advisors and supervisors, this enterprise is changed into Limited Incorporation with share ownership of 60% for village government, 35% for the third party, and 5% for villagers.

Considering the result of observation using communicative ethnography method, it can be seen that Panggungharjo village uses Javanese culture in managing village enterprise including

Table 4. The analysis of local wisdom-based BUMDes' Entrepreneurial Behavior.

No	Category	Findings
1.	Resource Utilization	The natural resource includes the utilization of social, economic, cultural, and technological landscape. Exploring from the social landscape related to the social condition of society such as RPS (Rubbish Processing House), Tamanu oil and used oil processing constituting a technological landscape because it uses technological creativity approach. Cultural landscape in Kampoeng Mataram included the application and the delivery of understanding on society life in Panggungharjo Village, related to the high appreciation for art and culture.
2.	Creating innovation	Innovation made in Kampoeng Mataram includes the products sold in the food stall, in addition to food, for example a variety of food menus, selfies (self-photograph taking) places such as photobooth, flowers, plant seedling need, and new building placement.
	Innovation change Work Mechanism	Making innovation in RPS (Rubbish Processing House) to deal with overtime problem making many employees absent in their work due to fatigue and to reduce their workload includes two alternative solutions: whether through elevate and to level off the hangar in order to be in parallel with dump truck so that the rubbish sorter can go up to it to unload the rubbish or through using conveyor.
3.	Dominant in the market	Considering the sufficiently significant price of tamanu oil in the market and the ownership of technology and market, this product is determined to be a superior product. This BUMDes is the only one producing tamanu oil in Indonesia. The superior product should qualify certain requirements in the term of raw material and strong funding. Therefore, it takes initiative to change its business form into Limited Incorporation in order to access external funding.
4.	Local Value	"Memayu Hayuning Bawono", "Memayu" means preserving or conserving, "Hayuning" means beauty or achieving prosperous or *rahayu* "well-being" condition, so that if we conserve the nature, it will, of course, exert a positive effect on the society's social life.
5.	Local Knowledge	Kampoeng Mataram business unit tries to explore the ancestor heritage of not using msg (royco) by reappearing local culinary culture using *Tempe bosok* (putrid *tempe*). It attempts to recover the local content of wearing Javanese fashion, using Javanese building, changing the guest's mindset about rural circumstance, recovering and reselling the childhood memory.
6.	Local Skill	Including the ways and the stages of cultivating farmland, the ability to cook naturally without course, for example, the 50-year old women actually have traditional culinary content and ancient heritage used to be sold in Kampoeng Mataram.

life guidelines, local value, local knowledge, social interaction, social rule, ethic, and language. Local wisdom the BUMDes has is the tenet "*Memayu Hayuning Bawono*". "Memayu" means preserving, "Hayuning" means beauty or achieving prosperous or *rahayu* "well-being" condition, so that if we conserve the nature, it will, of course, exert a positive effect on the society's social life. This tenet educates environment-friendly farming, not dependent on chemical fertilizers. Recovering the value "Hamemayu Hayuning Bawono" means conserving the earth without contaminating it by leaving chemicals on it. This is the local ability to enable farmers to perform environment-friendly farming.

The wellbeing approach the BUMDes Panggungharjo takes not only concerns the economic factor to remove the worry that all resources will be exploited so that they will be damaged but also emphasizes on natural conservation, cultural conservation of "Hayuning Bawono". The improvement of economic quality is not the main goal, but the improvement of economic quality will result from the improvement of natural and social-cultural qualities. Thus, the economy is only the impact of what has been performed corresponding to the tenet "Memayu Hayuning Bawono". It is attempted to be emphasized on so that at a certain point, the disciplinary understanding will appear, and then the members' affection awareness and care will be getting better.

4 CONCLUSIONS

Communication practice in BUMDes organization is useful to open a communication space to the employees in order to have an idea and thought, to discuss the improvement of production capacity, packaging, and income, and the difficulty and problem encountered by business unit to get a solution to them. It is also useful to explain the objective, program, work target, and to make the employees getting closer to each other so that the structural relationship with the management will not be rigid. It helps the employees in other division deliver information about work schedule, event agenda, and work an hour and facilitates the work coordination between different business units.

Organizational culture in BUMDes Panggungharjo for sustainable rural entrepreneurship management based on local wisdom can be seen from social ritual performance in the form of interaction and behavior of organization's members in discussing the problem to follow up the consumers' problem immediately. The management has a sensitivity to the employees' grievance and the employees remind each other, evaluate the work, motivate, encourage, and fight in BUMDes organization. Informal language "*kekancan* (friendship)" is used to facilitate communication between superior and subordinate. The manager emphasizes that anything should always be initiated by the top manager, suggestion and critique can be given in a bottom-up manner. All employees are involved in communication to make innovation for the sake of business sustainability. Director of BUMDes does not limit but explores innovation from the bottom to build the employees' loyalty and contribution to the company.

Ritual performance of an organization is an important appearance involving all workgroups with several regularities so that there are job distributions such as manager that is strategic in nature, program development, and relation with outsiders. Furthermore, it is the unit head that deals with the technical problem and the coordinator of division that conducts daily field supervision synchronized with both employees and guests. Director of BUMDes emphasizes that employees should be able to achieve work target that can impact on income, to understand and to maintain the working rhythm, to look for profit and operational efficiency in respective business units.

Local wisdom the BUMDes of Panggungharjo Village has to manage sustainable rural entrepreneurship is the utilization of resource in the form of social, economic, cultural, and technological landscapes. Local knowledge is defined as re-exploring the ancestor's heritage by reviving local culinary culture, recovering local content by means of wearing Javanese clothing, using Javanese building, creating a rural circumstance, retrieving childhood memory in rural tourism. Local skill includes ability, technique, and stage (procedure) in cultivating rice farmland and ability of cooking naturally without taking any course containing the ancestor's traditional heritage culinary, recovering "Hamemayu Hayuning Bawono" value meaning conserving the earth without contaminating it by leaving chemicals on it to enable the farmers to operate environment-friendly farming.

The author recommends the Village Government and BUMDes of Panggungharjo Village to pay attention to the welfare of employees working in Rubbish Processing House (RPS), to facilitate the administrators of Rubbish Bank's need by providing rubbish transporting vehicle, getting raw material not dependent on outside area but by planting Tamanu seed themselves, dealing with the work envy growing among the employees of Kampoeng Mataram by strengthening sense of belonging and organizational values, providing reasonable facilities to street sellers (PKL) in Kampoeng Mataraman and giving the villagers the opportunity of being entrepreneur as broadly as possible.

REFERENCES

Agunggunanto, E.Y., Arianti, F., Kushartono, E.W., & Darwanto. 2016. Pengembangan Desa Mandiri Melalui Pengelolaan Badan Usaha Milik Desa (BUMDes). *Jurnal Dinamika Ekonomi dan Bisnis* 13(1): 67–81.

Andrew, K. 2015. The Challenges of Entrepreneurship As An Economic Force In Rural Development: A Case Study of Kyaddondo East Constituency, Wakiso District in Uganda. *East African Journal of Science and Technology* 5(1): 105–120.

Anggraeni, M.R.R.S. 2016. Peranan Badan Usaha Milik Desa (BUMDes) Pada Kesejahteraan Masyarakat Pedesaan. *Jurnal MODUS* 28(2): 155–167.

Creswell, J.W. 2009. *Research Design: Qualitative, Quantitative, and Mixed Methods Approaches*. 3rd ed. Thousand Oaks, CA: SAGE Publications.

Das, D.C. 2014. Prospects and Challenges of Rural Entrepreneurship Development in NER-A Study. *International Journal of Humanities & Social Science Studies* 1(3): 178–182.

Denzin, N.K. & Lincoln, Y.S. 2017. *The Sage Handbook of Qualitative Research*, 4th ed. Thousand Oaks: SAGE Publications.

Deshwal, S. 2015. Understanding The Youth For Embracing Rural Entrepreneurship as a Career. *International Journal of Applied Research* 1(13): 579–581.

Eko, S. 2014. *Desa Membangun Indonesia*. Yogyakarta: Forum Pengembangan Pembaharuan Desa (FPPD).

Gani, A.Y. 2014. *Understanding Entrepreneurship: Memahami Secara Cerdas Makna Entrepreneusrship yang Sebenarnya*. Malang: Universitas Brawijaya Press.

Hayyuna, R., Pratiwi, R.N., & Mindarti, L.I. 2014. Strategi Manajemen Aset BUMDes Dalam Rangka Meningkatkan Pendapatan Desa (Studi pada BUMDES di Desa Sekapuk, Kecamatan Ujungpangkah, Kabupaten Gresik). *Jurnal Administrasi Publik (JAP)* 2(1): 1–5.

Ife, J.W. 2002. *Community Development*. Melbourne: Longman.

Kashani, S.J., Mesbah, A., & Mahmoodi, S. 2015. Analysis of Barriers to Agricultural Entrepreneurship Development from the Perspective of Agricultural Entrepreneurs in Qazvin Province. *Journal of Applied Environmental and Biological Sciences* 5(12): 47–55.

Korsgaard, S. & Müller, S. 2015. Rural entrepreneurship or entrepreneurship in the rural – between place and Space. *International Journal of Entrepreneurial Behavior & Research* 21(1): 5–26.

Lassithiotaki, A. & Roubakou, A. 2014. Rural Women Cooperatives at Greece: A Retrospective. *Open Journal of Business and Management* 2: 127–137.

Littlejohn, S. W., Karen A. F., & Oetzel, J.G. 2017. *Theories of Human Communication*, Eleventh Edition. Long Grove, Illinois: Waveland Press, Inc.

Mayer, H., Habersetzer, A., & Meili, R. 2016. Rural–Urban Linkages and Sustainable Regional Development: The Role of Entrepreneurs in Linking Peripheries and Centers. *Journal of Sustainability* 8:1–13.

Munyanyiwa, T. & Mutsau, M. 2015. Rural Entrepreneurship Challenges and Opportunities: A Case Study of Uzumba Rural Area in Zimbabwe. *International Journal of Multidisciplinary Research Hub* 2 (7): 15–20.

Nguyen, C., Frederick, H., & Nguyen, H. 2014. Female entrepreneurship in rural Vietnam: an exploratory study. *International Journal of Gender and Entrepreneurship* 6(1): 50–67.

Pacanowsky, M.E. & Trujillo, N.O. 1982. Communication and Organizational Cultures. *The Western Journal of Speech Communication* 46(2): 115–130.

Pato, M.L. & Teixeira, A.A.C. 2014. Twenty Years of Rural Entrepreneurship: A Bibliometric Survey. *Journal of Sociologia Ruralis* 56(1): 1–26.

Patton, M.Q. 2002. *Qualitative Research and Evaluation Methods*. Thousand Oaks: SAGE Publications.
Phungwayo, L.G. & Mogashoa, T. 2014. The Role of Entrepreneurship on the Socio-Economic Development of Rural Women: A Case Study of Kwa-Mhlanga in the Mpumalanga Province (Republic of South Africa). *International Journal of Business and Social Science* 5(9): 71–77.
Safitri, F.A., Susilowaty, E., & Mahmudah, S. 2016. Tinjauan Yuridis Terhadap Pengelolaan dan Pertanggungjawaban BUMDES Yang Belum Berbadan Hukum. *Jurnal Diponegoro Law Review* 5(2): 1–17.
Taghibeygi, M., Sharafi, L., & Khosravipour, B. 2015. Identifying factors influencing the development of rural entrepreneurship from the perspective of farmers of West Islamabad country. *Research Journal of Fisheries and Hydrobiology* 10(10): 161–168.
Toosi, R., Jamshidi, A., & Taghdisi, A. 2015. Affecting Factors in Rural Entrepreneurship. (Case Study: Rural Areas of Minodasht County). *Journal of Research and Rural Planning* 3(8): 1–3.
West, R. & Turner, L. H. 2013. *Introducing Communication Theory: Analysis And Application*. 5th Edition. New York City: McGraw-Hill Education.
Yin, R.K. 2009. *Case Study Research Design and Methods*, 3rd ed. Thousand Oaks: SAGE Publications.

Rural Socio-Economic Transformation – Kinseng et al. (Eds)
© 2019 Taylor & Francis Group, London, ISBN 978-0-367-23603-8

The communication characteristics of local elites in development programs: A study case in Pandeglang, Banten, Indonesia

N. Fitriyah
Department of Communication Science, Faculty of Social and Political Sciences, Sultan Ageng Tirtayasa University, Banten, Indonesia

S. Sarwoprasodjo, S. Sjaf & E. Soetarto
Department of Communication and Community Development Sciences, Faculty of Human Ecology, Bogor Agricultural University, Bogor, West Java, Indonesia

ABSTRACT: Pandeglang is one of the districts with a high poverty rate in Banten Province. Local government has released a social security program since 2011 known as Jaminan Sosial Masyarakat Banten Bersatu. The communication characteristic of the Pandeglang's citizens are intervened by Moslem leaders or known as "ulama". Patrimonial system and patron client culture are strengthening the position of ulama as the local elite in contributing to the district's development. Therefore, ulama is seen as a credible source to communicate the messages of development and empowerment to the community. A qualitative approach with a single case study method is used where interview, observation were conducted as data collection technique. As the results, ulama and its religious institution is seen as the agent of change and innovator, who provide critical awareness to empower the people in poverty.

1 INTRODUCTION

1.1 *Background*

Since Banten became a province in 2000, especially in Pandeglang Regency, the problem of poverty is the obstacle in the development. The government's attention to the problem of poverty was outlined in the form of Jamsosratu in 2011. Jamsosratu is a social security for the people of Banten, where fund is delivered to Rumah Tangga Miskin (RTSM or Poor Households) especially to pregnant women, breastfeeding women, and family with children range from 7–18 years old. This assistance is aimed to improve social functioning and empowerment.

All regencies and cities in this Banten province received the Jamsosratu policy, especially Pandeglang District, which has the highest poverty rate. The efforts made by the Pandeglang district government in alleviating poverty provide great hope for the community to escape poverty, however, the program haven't showed significant results. The poverty rate keeps growing, despite the enactment of Jamsosratu program. The following data are obtained from the Central Bureau of Social Statistics, Pandeglang Regency.

Table 1. The poverty level of the population of Pandeglang Regency (2013–2017).

Year	2013	2014	2015	2016	2017
RTSM	505	862	15,157	15,157	9956
Poverty	121.45	113.14	124.42	115.9	501.97

Source: The central agency of the Pandeglang District statistics 2017.

Several studies had identified some problems in Banten Province to decrease the poverty rate. The implementation of the Jamsosratu policy in Banten Province has not been carried out according to the policy principles and did not give any impact on people's welfare (Hazrumy 2016). A comprehensive evaluation requires the implementation of the Jamsosratu program, so that Jamsosratu has a significant impact on poverty management (Gustaman 2016). Several other problems in poverty reduction lead to poverty alleviation which tends to be charitable instead of empowering, so that people do not have the initiative to be independent (Prawoto 2008). This finding is reinforced by the illustration that poverty reduction should be based on the integration of cultural, religious and patterns of relations between the government and the community (Rozuli 2010). These findings provide a lesson that poverty reduction can no longer be done by top down methods. Some other findings, also assert that socio-cultural forces that put forward the principles of local culture have to be an important element in poverty reduction (Kifli 2016).

Departing from the findings above and the sociological conditions of Pandeglang Regency known as the city of a thousand ulama with a million santri, where the ulama is one of the important element for development. Ulama have high influence in determining people's attitudes and behavior. This is not only because of their political position but more because of their authority, charisma, inherent myths, or because of their knowledge and experience (Karomani 2008). The special position of the ulama made the Pandeglang community tend to submit and represent their interests and aspirations to the ulama so that the community in general would listen to the direction and opinions expressed by the ulama. Ulama as community leaders and community role models have a significant role in assisting development programs by providing a religious foundation to see community development and empowerment as a part of worship (Fajri & Wicaksono 2017).

The social role of ulama has been tested in shaping the social order of society (Muslim et al. 2015). Communication and social interaction carried out by ulama are characterized by their desire and determination to make the community more empowered. Informal activities such as routine recitation of Qur'an or known as 'Majlis taklim' and meetings in the village environment and deliberations at the district level, ulama convey their views on how development programs can have a significant impact for the welfare of the community. In daily informal activities, contexts lead to how the community has the initiative to be free from poverty.

Assisted by Jamsosratu program and strong believes to ulama, Pandeglang community still needs more motivation to free themselves from poverty. The Pandeglang community assume that Jamsosratu program as an obligation from the government to free the community from poverty. Yet, the program objectives are not limited to give fund assistance the community but also to empower them so that they do not have to depend on government's assistance. This phenomenon makes the ulama feel called to straighten out what the ulama think about the Jamsosratu program.

The main proposition in this study is that communication between ulama is very difficult to reach consensus, which leads to a problem where the Jamsosratu program is not fully delivered and understood by the community. This indicates that the communication practices of ulama in development has failed to serve its purpose. The discussions, arguments and maneuvers carried out by the ulama indicated that the communication of the ulama is not legitimate for the community, because there is discrepancy in perception between ulama, the community and stakeholders about the Jamsosratu program.

The stretch of communication problems experienced by ulama in the Jamsosratu program can be seen from the failure of ulama to build a valid and proper message in explaining Jamsosratu program. The communication problems experienced by the ulama in the Jamsosratu program can be seen from the failure of the ulama to build a valid and appropriate message in explaining the Jamsosratu program. On the other hand improving the message program by continuing to explain that the Jamsosratu program is a program that is truly intended to improve the quality of people's lives not for others. However, the community cannot decide the purpose and objectives of the program, thus giving rise to miscommunication which means the message cannot be received correctly by the community. Ulama, as a local elite, have been proven in building public awareness through communication practices. However, it is also necessary to study how the communication characteristic of ulama and its role in supporting development programs. Thus, the objectives of this study are describing the communication characteristic of ulama in the Jamsosratu program and explaining the role of ulama in empowerment and supporting the Jamsosratu program.

1.2 *Novelty*

This study seeks to discover the implications of development communication problems in Pandeglang Regency. The development communication problems in Pandeglang is originated from the social, political and cultural systems in Pandeglang which includes the participation of informal leaders and local elites, such as ulama. The unbalanced power relations in local political practices in Pandeglang have inhibit the regency's development which results to social marginalization. Research with a focus on the communication characteristics of ulama in development programs explores how messages about development program are delivered to the community. The process of messages delivery includes interaction between community and ulama in order to produce collective or mutual understanding. Here, mutual understanding is seen as a legitimate force so that the ulama and community have important roles in developing and delivering messages about development programs.

Based on assumptions above, this study offers new approaches which are: (1) the construction of communication characteristics, interactions and relations of ulama who have roles in empowering community; (2) communicative actions need to be based on local structure and the culture of community; and (3) balancing the need of rationality and objectivity in implementing development communication program with the need to recognize the socio-cultural factors that might be not rational nor objective.

2 METHOD

2.1 *Research design*

Qualitative approach is used in this study. Qualitative research used variety of methodologies and practices, which have different meanings at each stage of their development (Denzin & Lincoln 2009). The author then study the social settings and communication settings of the ulama to understand and interpret the Jamsosratu program. The author then chose the case study method as a basis for understanding the communication characteristics of ulama based on differences in values, beliefs and scientific theory.

2.2 *Sampling*

The study was conducted for six months, from July to December 2018. The data collection technique was done purposively where key informants or people who are knowledegable in Jamsosratu program. The key informants are ulama, bureaucrats, village community leaders and ordinary people.

Table 2. Mapping research informants.

Key informants	Key informants (micro)		Informants
(macro)	Sub-district	Village	
6 Ulama 1 Bureaucrats	Menes	Kadupayung 3 Ulama 1 Community leader 1 Bureaucrat 2 Grantees	10 Person
1 Academician 1 Program practitioner	Saketi	Bojong 3Ulama 1 Community leader 1 Bureautcrat 2 Grantees	
Total: 22 key informants			

2.3 *Population and sample*

The research was conducted in two sub-districts that could represent the problems of study. The population in this study is the whole community of Pandeglang Regency while the sample is the parties involved, have active involvement and are concerned about Jamsosratu program and development. To make it easier to get data, researchers then categorize informants into two, namely (1) key informants and (2) informants. Key informants are informants who are directly related and intensively related to Jamsosratu development and programs, while informants are scattered informants, having indirect links to Jamsosratu program.

From this population, the researcher then divides into two parts: macro level and micro level. Macro level is aimed at data related to the involvement of ulama in the Jamsosratu program and development at the district or provincial level. Micro level is aimed at data from several key informants whose activities are within the sub-district and village scope. A purposive method is used in determining the key informants and informants.

Observations and in-depth interviews for macro level key informants were carried out on the activities of informants who led formal institutions at the District and Provincial levels such as the Indonesian Ulama Council (MUI), Muhammadiyah, Matlaul Anwar (MA), Nahdatul Ulama (NU). The forums observed were formal and informal meeting forums held regularly at the institution. Observations and interviews at the micro level were carried out in formal and informal forums of citizens, such as the village meeting forum, Musrembang, routine recitation of Majilis taklim and community gatherings.

2.4 *Data collection*

In qualitative research, data explain the phenomenon of problems in the field. The primary data is collected through in-depth interviews and participant observations. Meanwhile, secondary data are obtained through literature study techniques as complementary data to confrim the primary data.

In-depth interviews were conducted to key informants to explore their personal experiences in the Jamsosratu program and development issues. The data sought in this interview relates to communication characteristics and the role of ulama in development which are divided into several dimensions: (1) communication dimensions which consist of communicator, messages, media, communicator, and effect variables; (2) dialogue dimensions which consist of motives, situation, argumentation, participant status, relations and appreciation of participants; (3) the role dimension consists of participation, arena of activity, interaction patterns and opinions developed; and (4) the public space dimensions which consist of models of public space, goals, functions, arenas, opinions, goals, and interactions.

The observations were conducted by visiting the object of the case study for six months from July to December 2018. Observation involves observing meetings, communication activities, development activities and others. Participation observations were carried out to observe how political communication was carried out by ulama in the Jamsosratu program and development. The observations were also conducted as a part of data triangulation, to clarify and verify data.

2.5 *Data analysis*

Creswell (2012) explains that data analysis is a process where there is continuous reflection of the data obtained. Referring to this, data related to communication characteristics and the role of ulama in development are analyzed through stages: (1) carefully reading all important notes of research (2) providing codes on important topics, such as communication dimensions that include messages, media, motives, the arguments and effects of ulama communication (3) compiling typologies related to communication characteristics and (4) reading literature related to the communication characteristics of ulama. Based on the results of the coding above and after all were analyzed, the researcher then reconstructed in the form of descriptions, narratives and arguments.

3 RESULTS AND DISCUSSION

3.1 *The communication characteristics of ulama*

Pandeglang community is a community that strongly abide to its culture, costums, and tradition. This special characteristic has implications for the communication process and patterns of interaction between ulama. 'Ulama' in Arabic word meaning 'a scientist or researcher'. In Indonesian, however, the meaning of ulama is 'people who are experts in Islamic studies' (Nasution 2018).

The process of communication between ulama-community-government in the Jamsosratu program is a process of interaction that has a purpose. It does not take place in a vacuum, but it is always related to the context, the particular situation and the space where the communication takes place. Generally, the context of communication that takes place between ulama-community-government is more directed at psychological aspects (attitudes, tendencies, prejudices, and emotions) and social aspects (norms, social values, cultural characteristics). These contexts provide their own features and variants in its communication practices. The communication practices and arguments presented by the ulama are generally shown to build discourse, control and criticism of the Jamsosratu program implementation practices.

When there was a formal meeting of residents about the Jamsosratu program, the communication practices reflected the psychological condition of the community who had high hope of getting the Jamsosratu program grant. This phenomenon was seen when the community began to convey their living conditions that were still in the category of Very Poor Households (RTSM) and needed assistance from the government. In this situation, the communication practices of the ulama is to motivate the people to improve themselves and to become closer to God.

We found it interesing that ulama have different interest. A group of ulama tried to encourage critical thinking yet another group of ulama is trapped in pragmatic political issues. These contradiction leads to different messages delivered to public: a critical thinking for empowerment and development and a reason for ulama to gain personal interest.

3.1.1 *The communication characteristics of traditional and modern ulama*

There are two typologies of ulama's communication practices. Ulama are divided into two, namely "traditional" ulama and "modern" ulama. Traditional ulama is a group ulama who maintain their traditional role as religious leaders and have boarding schools and lead Majelis Taklim, while modern ulama is a group of ulama who engage in the politics. Each of them have its respective characteristics and communication content, even though there are still similaraties between them.

Traditional ulama have decent knowledge and views, especially in the Jamsosratu program, to solve the social problems occured in the community. Tradition ulama has been delivering "building critical awareness" messages to the community so that the community is motivated to empower themselves. The message delivered by traditional ulama is also comprehensive thus have impact to the welfare of community. The traditional ulama also created a public space where community could have dialogue about problems faced by community. Generally, the traditional ulama have become initiators and mediators in the realization of public welfare.

Regarding the Jamsosratu program, the communication practices of traditional ulama are not only limited to their cultural and structural strengths, but also formed through conversation and arguments that are seen in formal and informal dialogues. The dialogue between ulama, community, and government from perspective of communicative action is targeted to achieve consensus not only in small scale but also large scale such as at Musyawarah Perencanaan Pembangunan Desa (Musrembang), a community discussion forum that discuss about Jamsosratu development programs.

"Modern" ulama is a group of ulama who engage in practical politics. The communication characteristic of modern ulama consist of three type of messages: (1) ulama conveyed political messages instead of empowerment messages; (2) ulama conveyed stronger interventions that weaken the cultural social order and marginalize society in development and (3) the messages are co-opted by political interests in the name of development. Based on this findings, poverty in Pandeglang Regency is caused and maintained the unequal social relations between ulama-government-society.

Modern ulama, however, have another communicative competence, which is negotiating and building consensus with politic stakeholders so that ulama can have access to political parties and development programs. This competence makes ulama powerful enough to penetrate community and government institutions. The power gained than used by ulama to be local elites that further strengthened the local elitism in implementing development programs in Pandeglang.

3.1.2 *The communication process of traditional and modern ulama*

Habermas (1987) stated that communicative action is a communication process that prioritize rationality and comprehensiveness. The ulama, in communicating development programs, use the communicative action process which includes rationality in building and delivering messages about development. However, there are differences regarding the arguments, goals, and impacts delivered. Traditional ulama use high context communication with a patron-client approach. The patron-client approach, however, makes the communication less rational since the ulama use his power to influence the community perception about development programs. On the other hand, modern ulama use low context communication so it is easier for community to understand messages about development programs.

Communication of ulama in the Jamsosratu program is closely related to communication networks established by ulama with the community and government. Building network is seen as communicative action to reach the interest of ulama so that they could have more influence to produce valid and legitimate messages about development. The traditional ulama is face with difficulties in building the critical awareness of community since they do not have political power and use high communication context. Yet, the modern ulama is easier and faster in reaching concensus since they have political and economic power. This is closely related with the previous findings which stated that traditional ulama is more aware of the empowerment of community while the modern ulama tends to deliver their political interest.

In interpersonal communication, for example, traditional and modern ulama have discussion with directly with members of community about Jamsosratu program. They discuss about the polemic of the number of recipients, procedures, disbursement processes, and the monitoring and evaluation of the program. This type of communication allows the community to directly process the verbal and non verbal cues of ulama's communication process. To be specific, the communication practices represent a dyadic communication where ulama communicate directly and personally with the grant recipients, government apparatus, or facilitators of Jamsosratu program. For example, when KH Bai from Menes Subdistrict tried to calm down a citizen categorized as RTSM communities but is not listed as recipients of Jamsosratu program:

> *'Citizen: Pak Haji (a nickname for Ulama), what to do, I'm a poor person and have no permanent job. Even if I took several part time jobs, I won't meet the need of my family. I have three children and a pregnant wife. I have reported my sitution to local elits so that I could be the recipient of Jamsosratu program, but there's no update until now. Can you help me so that I become the recipient of this program? I beg you, Pak Haji.*
>
> *Ulama: You have to be patient in this situation. I will try to talk to government, but you have to work hard and be sincere in life. Remember that you have to give your full trust to Allah (God) in order to survive'.*

The direct communication between citizen and ulama is often encountered in the field. This type of communication is rarely found in bigger meetings where several recipients and other stakeholders participate. In bigger meetings, group communication is practiced in order to reach a mutual understanding about Jamsosratu program. In group communication, the interaction is more dynamic, involved more participants such as facilitators and the recipients. However, when bigger meetings occured, interpersonal communication also occured since it is the opportunity for citizen to meet directly with ulama.

Variety of opinions arises in group communication. Usually, a group communication only discuss about the problems of Jamsosratu program. In this situation, interpersonal communication also occured since the communication of ulama could be seen, where they try to deliver arguments about Jamsosratu program so that their arguments could be seen as legitimate by community.

Referring to the communication patterns of ulama in development, there are several distinctive characteristics which are: (1) the socio-cultural factors which are religious with patron-client relationship make the ulama use high context communication; (2) openness, ulama try to be open and honest; (3) empathy where ulama try to understand the motivation, experiences, emotion, attitude, and hope of the community to Jamsosratu program; (4) equality, where ulama subtlely respect the community and gives opportunity to the communty to communicate with them.

These findings showed that communication between ulama and the community is a normal occurence in Pandeglang. Yet, this communication support the community development since ulama gave influence to the perception of community about development programs. It is interesting because based on patron-client relationship, ulama is able to build a valid and legitimate message that could affect the development of Pandeglang. In a democratic society, a valid and legitimate message have to be produced by constitutions that founded on society's sovereignty. In Pandeglang, technocracy and patron-clientism are practiced so it is hard to produce a valid and legitimate message about development. This condition further marginalized community and do not bring any critical awareness to the community.

3.2 *The role of ulama in empowerment*

Empowerment is described as a process and how a person can be empowered. Empowerment is also defined as the process of transitioning from a state of helplessness to a state of relative control over one's life, destiny, and environment. Community empowerment is an effort to restore or enhance the ability of the community to be able to do according to their dignity and dignity in carrying out their rights and responsibilities as a community (Mubarak 2010).

Some literature shows that community empowerment have to be based on the sociological conditions of the local community so it is easier for the people to understand the empowerment programs. The existence and independence of the community comes from internal strength and is not destructive (Ningrum 2012). The sociological strength of the community in empowering one of them lies in the role of community leaders. The involvement of community leaders in empowerment programs is very effective in changing people's behavior (Satriani et al. 2015). Development is supposed to involve community leaders and religious leaders or ulama from the planning process to monitoring development (Siagian 2016). Ulama as community leaders in the empowerment program incorporates modern cultural values such as hard work, frugal behavior, and openness to society (Alam 2016).

The main responsibility in the development program including the poverty program is that the community is empowered or has the power, strength or ability. The ability of the ability in question is cognitive, conative, psychomotor and affective abilities and other physical/material resources (Widjajanti, K. 2011). In practice empowerment involves four aspects: cognitive, psychomotor, psychological, economic and political aspects. The main priority of empowerment programs is that the community must be empowered, have the strength, and have the skills/abilities to survive and improve the quality of life.

In the Jamsosratu program, ulama are involved in the empowerment process, even though it is informal and non-structuralist. Yet, ulama is able to build discourse and reach concensus so that the community understand the objectives of Jamsosratu program. Ulama can act as mediator when different opinions about program arises and ulama can also act as innovator because they can solve development problems. The innovation stated in this study is referred to the use of religious discourse to motivate the participation of community in development programs.

Particularly on discussion about empowerment, ulama communicate three aspects at the micro level which are cognitive, conative and affective. This type discussion in formal and informal forums, such as community gatherings, urun rembug and Majelis Taklim, or a day-to-day personal meeting between ulama and community. For cognitive aspect, ulama try to communicate messages that motivate the community to have critical awareness. Ulama state that community need to be knowledgable so the community could be empowered, one of the example stated by ulama is "The objective of Jamsosratu program is to empower the community, not to drive the community become consumptive". For conative aspect, ulama focus on attitude and behaviour of community so that they are more aware to development and empowerment. While for affective aspect, ulama focus on the effort to motivate the community and local government to have more awareness on how to empower the community.

Empowerment efforts carried out by ulama in these three aspects (cognitive, conative and affective) contribute to the creation of community independence. Thus, the people of Pandeglang Regency through the Jamsosratu program have sufficient insights reinforced by the desire to participate in development. The empowerment process through Majelis Taklim, and through formal informal forums conducted by ulama gradually lead the community to have knowledge so that the community with the help of the Jamsosratu program can improve the quality of life.

The empowerment process carried out by ulama in formal and informal forums, both at the macro level and micro level, represents a significant influence to development programs. The discussion about the Jamsosratu program that was delivered to the community was an effort to improve the capacity of the community. Communication content that is conveyed to the community inspires cognitive, affective and conative aspects so that society gradually become independent and can be free from poverty.

At macro level, ulama communicate different issues. Since at the macro level ulama have to communicate with government, so ulama try to build government awareness in planning and creating empowerment and development program. The ulama, which is united in MUI, Muhammadiyah, and NU, express their aspirations on how development programs have to be focused on poverty alleviation programs and improving the education for community, given the high poverty rate in Pandeglang.

Ulama who have major roles are modern ulama since they have access, arena, ability, and authority to be able to communicate with government. Ulama state to the government that empowerment and development programs could be planned or discussed through Majelis Taklim, where community gathers and can be mobilized.

> The Majelis Taklim is a routine community meeting which functions as a place of Qur'an recitation but also functions as a community gathering facility. In this Majelis Taklim the community came voluntarily to meet, discuss and exchange ideas with other communities, especially with the ulama. Of course it would be more efficient and effective if the Majelis Taklim also functioned as a public space to empower the community, while still promoting the principles of religion and local norms.

Ulama view that Majelis Taklim is the easiest and accessible space to build the critical awareness of the community, so that the cognitive aspects of society can be directed towards empowerment efforts. Ulama could act as a social control so that there are no miscommunication about Jamsosratu program. A clear message about Jamsosratu program could reach the objective of Jamsosratu program. The social control is carried out through informal meetings both when dyadic communication or group communication are practiced.

At the micro level, traditional ulama have important roles in managing Majelis Taklim at the village level, while interacting and mingling directly with the community. Through this opportunity, tradition ulama try to explain the objectives of Jamsosratu program and wish that Jamsosratu program could empower the community. The message of Jamsosratu program is based on religous foundation, where people have to work hard in order to have a good life as stated on verses in Al Qur'an.

Based on the results of observations, there are several communication events where traditional ulama communicate with the community. The first type of communication events is through dialogue. Dialogue is a conversation between ulama and the community to reach a mutual understanding and a sense of calmness in pursuing people's needs in life. The effort to reach a mutual understanding is conducted through the development of mutual respect between ulama anda community. This effort is reflected when ulama and community sat together in a community meeting which discuss about Jamsosratu program. The second type is through sharing. Sharing is a means for ulama to exchange opinions and share experiences. Here, ulama, community, and local government share about their experiences in Jamsosratu program. They usually discuss about problems in the implementation of Jamsosratu program, where list of recepients have to be updated and the whole process of accepting the assistance is hard to be done by the community.

The third type of communication events is counseling. Counseling is used to clarify the problems faced by the community and or where community ask ulama to become a mediator to solve the problem. Regarding the implementation of Jamsosratu program, counselling only happened at certain time, which is very incidental and very personal. Counseling that has been observed is very informal and takes place at the residence of ulama. While not all of citizen or local government apparatus have enough courage to ask ulama for counselling, but ulama is always open to everyone when they need opinion or solution.

4 CONCLUSION

The interaction between ulama, the community, government, and other stakeholders is a process of communication that has a purpose and does not take place in a vacuum. It is related to the socio-cultural conditions and psychological conditions of the community. The communication characteristics of ulama is divided into two typologies, namely: traditional ulama and modern ulama. In general, the communication built by traditional ulama and modern ulama contains elements of comprehensiveness and is able to reach consensus, which is the creation of public space so that people can dialogue about existing problems.

Specifically the communication characteristic of traditional ulama used a patron client approach, where citizens is seen as client of ulama rather than rational relationships. On the other hand, the communication characteristic of modern ulama use a high context that requires additional information to understand the meaning of the content or message of communication. The communication characteristic of modern ulama also strengthened the identity of ulama who are powerful enough to penetrate the community and/or government institutions.

Both the traditional and modern ulama act as agents of change and innovators. The influence of ulama is covered from macro to micro levels. However, there are differences between traditional and modern ulama: traditional ulama tends to blend with the community, while modern ulama tends to be more institutional to strengthen their political power. The message of community empowerment is delivered by ulama through three forms, which are dialogue, sharing, and counseling.

There are some interesting findings in this study which illustrate the other side of the communication characteristics of ulama in the Jamsosratu program, namely "communication map" of scholars. The communication map is intended as the realm of strength and the arena of communication of scholars. Communication activities in the Jamsosratu program at the micro level are dominated by traditional ulama. The more intensive traditional ulema provides understanding and encourages critical awareness of the community in order to be free from poverty through the Jamsosratu program. Traditional ulama are more often present in formal and informal meetings of the village community, and in these meetings messages about empowerment are delivered. The communication intensity of traditional ulama indirectly makes a strong emotional bond between the community and the ulama at the micro level, so that this is why traditional scholars have more influence and their existence is rooted in the midst of the village community. While modern scholars focus more on (1) strengthening institutional capacity (2) efforts to encourage development policies at the macro level (3) involved in pragmatic political issues both at the district level and province. Modern scholars consider that to change the condition of society must go through efforts in fighting for pro-community policies at the macro level.

Based on the findings, there are several recommendations to support the development programs in Pandeglang: first, the participation of ulama in development program, especially in

Jamsosratu program, should be formalized so the stakeholders can facilitate coordination between ulama and people. The stakeholder also need to consider that ulama have an important role, not only supporting role, in implementing development programs. Secondly, the Jamsosratu program have to be enacted with other approaches to encourage the empowerment of the community's economy through micro-business financing, so that the community could be independent economically. Third, optimizing the use of public space by all stakeholders so that the issue of development can be discussed in a space where the community is free to express its aspirations equally without any pressure from any party.

REFERENCES

Alam, U.S. 2016. Indonesian challenges and opportunities in local community empowerment. *Government: Journal of Governmental Science* 1(1): 25–34.

Creswell, J.W. 2012. Research design qualitative, quantitative and mixed approaches. Yogyakarta: Student Library.

Denzin, N.K. & Lincoln, Y.S. (eds.). 2009. *Handbook of qualitative research* (translated). Jogjakarta: Student Library.

Fajri, A. & Wicaksono, B. 2017. The role of informal leaders in rural development (Study in Pulau Terap village, Kuok District, Kampar District, 2014). *Online Journal of the University of Riau's Faculty of Social and Political Sciences Students* 4(1): 1–10.

Gustaman, E. 2016. *Evaluation of Banten united people's social security policy Jamsosratu in Banten Province*. (Doctoral dissertation). Bandung: Universitas Pasundan.

Hazrumy, A. 2016. Implementation of Banten bersatu people's social security policy (Jamsosratu) in Banten Province. Dissertation. Bandung: Universitas Pasundan.

Habermas. 1987. *The theory of communicative action vol 2: Lifeword and system: a critique of functionalist reason*. Boston: Beacon Press.

Karomani, K. 2009. Ulama, jawara, and umaro: A Study of local elite in Banten. *Sosio-humaniora* 11(2): 168–172.

Kifli, G.C. 2016. Communication strategy for agricultural development in Dayak communities in West Kalimantan. *Agro Economic Research Forum* 25(2): 117–125.

Mubarak, Z. 2010. Evaluation of community empowerment was reviewed from the development process at the PNPM Mandiri Urban activity in the Sastrodirjan village of Pekalongan Regency. Dissertation. Semarang: Universitas Diponegoro.

Mulyana, D. 2011. *Communication science*. Bandung: PT. Rosdakarya.

Muslim, A., Kolopaking, L.M., Dharmawan, A.H. & Soetarto, E., 2015. Dynamics of the role of social politics of ulama and jawara in Pandeglang Banten. *Mimbar, Social Journal and Development* 31(2): 461–474.

Nasution, N.H. 2018. Da'wah Ulama Communication in South Sumatra. *Journal of Islam-ic Communication and Public Relations (JKPI)* 2 (1): 16–51.

Ningrum, E. 2012. Dynamics of traditional Kampung Naga community in Tasikmalaya regency. *Mimbar, Social and Development Journal* 28 (1): 47–54.

Prawoto, N. 2008. Understanding poverty and its coping strategies. *Journal of Economics & Development Studies* 9 (1): 56–68.

Rozuli, A.I. 2010. The diversity of religious and cultural institutions and their implications for strengthening the economic activities of rural communities. *Interactive: Journal of Social Sciences* 1(2).

Satriani, I., Muljono, P. & Lumintang, R.W.E. 2015. Participatory communication in the family empowerment post program (Case Study in RW 05 Situgede Urban Village, West Bogor District, Bogor City). *Development Communication Journal* 9 (2).

Siagian, H.F. 2016. Patterns of the involvement and impact of communication the ulama prevent conflict in the community. *Kareba: Journal of Communication Sciences* 2(1): 41–54.

Widjajanti, K. 2011. Model of community empowerment. *Jurnal Ekonomi Pembangunan* 14(1): 15–27.

Rural Socio-Economic Transformation – Kinseng et al. (Eds)
© 2019 Taylor & Francis Group, London, ISBN 978-0-367-23603-8

The struggle of cantrang fishermen in Indonesia: A pseudo victory?

S. Sarwoprasodjo, A.U. Seminar, R.A. Kinseng & D.R. Hapsari
Department of Communication and Community Development Sciences, Bogor Agricultural University, Bogor, West Java, Indonesia

ABSTRACT: Indonesia's government has banned the use of cantrang by fishermen in Indonesia in 2015, which drove a massive protest rally by fishermen in Indonesia. Even though government provided aids for cantrang fishermen, best solution has not yet been achieved. The purpose of this study is to explain the use of narrative theory by fishermen organization and social movement in Indonesia as a communication strategy in building rationality about the legalization of cantrang. Using qualitative methodology, 25 members of fishermen social movement organizations from four district (Indramayu, Tegal, Pati, and Lamongan) throughout West Java until East Java were interviewed. This study showed that fishermen social movement organizations used provocative narrative approach in their communication strategy to its members and to government, where social movement organizations blame the current Minister of Maritime and Fisheries Affairs to provoke a policy review on cantrang legalization.

1 INTRODUCTION

1.1 *Socio-economic transformation through Social Movement Organizations (SMOs)*

The rise of Social Movement Organizations (SMOs) since 1960s has lead many social scholars study its contribution to the society (Jennifer 2015). SMOs is a non-state actor that thrives to do socio-economic transformation, representing people who struggle to overrule hegemonic power. SMOs may be seen as venues of empowerment permitting those excluded from official public spheres to 'find the right voice or words to express their thoughts' (Fraser 1992). In short, socio-economic transformation needs two elements: (1) an inclusive arena to (2) communicate the words and voices of marginalized people, where these two elements are mobilized by SMOs.

In the study of Development Communication, SMOs is an actor that support a practice of communication for development and social change because they enable the people: (1) to hold their own communication to ensure their voices are heard; (2) to build horizontal dialogue with related stakeholders; and (3) to decide on issues that are closely related to their life for the sake of changes that benefit the people, not the hegemon (Gumucio-Dragon 2009). These shows how SMOs have to build their own strategy to make sure the voice of people are heard by the government and lead to a social change that benefit the people.

Many studies have used narrative theory to explain social movement strategy to communicate with authority. It is understood by social movement scholars that narrative is a powerful tool for collective action to achieve social change (Benford & Snow 2000, Polletta 2006). Narrative helps people to make sense of the world and in ways allow people to shape and form, interpret, and change the world around us. In short, narrative theory sees that human beings understand their lives in shape of stories. Narratives have an important role in consolidating memories, shaping emotions, and providing group distinctions, among other impacts (Houghton et al. 2013).

This paper seeks to explain the strategy of fishermen SMOs in Indonesia in building alternative narratives on 'cantrang', a local modified seine nets, that have been ban recently by the government because of its negative impact to environment. To persuade the government to lift the cantrang ban, the fishermen SMOs have been using provocative narrative so that transformation of socio-economic that benefit cantrang fishermen could be achieved. The theory used in this paper are narrative theory and Coordinated Management Meaning (CMM) to analyze the process of constructing provocative narrative about 'cantrang' in order to lift the cantrang ban by government.

1.2 *Narrative, provocation, and Coodinated Management Meaning (CMM)*

Many scholars, especially social movement scholars, are debating whether framing theory and narrative have similar functions. Polletta & Chen (2012) stated that narratives have to be persuasive, it has a to be logic and could influence people to agree with the story. Framing, while also derives from cultural narratives and and acts as a powerful rhetorical device like narratives, have limited understanding of how frames are shaped by audiences. Narratives derives from people who produce or act narratives: they know how to build a story, why they should tell stories, and how to respond to a story. Framing on the other hand defined by analyst rather than the actors of narratives. Narratives also have to construct a logical consistency as a criterion for persuasiveness. How do persuasion works to provide a better understanding?

The first step to understand how narrative could embody a persuasive character is by understanding that narrative is dynamic. Liu (2016) stated that narrative could be reconstructed since it is a contested spaces, where narrative seemed to have contradictions as they were unfolding. Narrative has beginning, a middle, and an end, and toward the ends is filled with conflicts, causal explanations, and the sequence of events (Steinmetz 1992). Narrative usually includes (1) time, plot, and progression; (2) setting, space, or perspective; (3) characters; and (4) reception or feedback (Herman et al. 2012). Characters in narrative include protagonist, antagonists, and bystanders. Events are recounted in order or sequence of events where structure emerge (plot). The end of the story is evaluative and projects a normative future, where people could take moral of the narratives (Polletta & Chen 2012).

One of persuassion mode in communication that usually used by activists, protesters, or social movements. The term provocation is understood as a positive stimulation, enhancing curiosity and critical thinking by many scholars from different disciplines or field, it is associated to violation of norms and to violence (Boudana & Segev, 2017a) or provocation act as a function in justifying one's actions and passing judgement over the opponent's actions, which includes a moral in its construction (Wahlstrom 2011). Boudana & Segev (2017b) stated that provocation is perceived as a tool to awaken people's consciousness regarding specific social or political issues.

Provocations could also be seen as struggles over symbolic power, as they attempt to overrule current standards and subvert authority and hegemonic positions or the drive social and political change (Driessens 2013). Definition from Driessens clearly support the rational of this paper that through provocative narration, SMOs are struggling to be heard by the government and to do a socio-economic transformation.

All of the previous studies mentioned above about narratives on social movements derives from sociology and politics. One of communication standing in explaining narrative is the provocation narrative that could help us in explaining how social movements construct their narrative through communication process. This paper, however, use theory of the Coordinated Management of Meaning (CMM) (Pearce & Pearce 2000) through a community dialogue process as a guide to explain how communication co-construct the narrative on cantrang among fishermen social movement organizations.

CMM sees communication a primary social process (Pearce & Pearce 2000). The communication perspective is grounded in the belief that mundane "talks" actually contribute to the meaning making process and relationship building. In short, the construct of social reality starts from communication. Pearce & Pearce (2000) further explained that: 'CMM locates each act simultaneously within a series of embedded contexts of stories about persons, relationships, episodes and within an unfinished sequence of co-constructed actions'. Every act is either preceed by previous act(s) and followed by subsequent act(s) where through this sequence of acts, the plots, characters, and relationship surrounding the characters are co-constructed. This paper seeks to explain the narratives of fishermen social movement organization as a communication strategy to legalize cantrang ban through exploring the sequence of acts proposed by CMM.

1.3 *Literature review on cantrang*

Discourse about cantrang is still considered new in Indonesia. The debate about the use of cantrang has only hit the media only because of the massive protest conducted by the fishermen SMOs in Indonesia. Thus, studies about cantrang are very limited. We have found studies about cantrang that only explore the techniques of using cantrang (Sasmita et al. 2012), environmental impact of cantrang (Wahyuningrat 2018), socio-economic of trawl (considered the same as cantrang) fisheries (Hendroyono 2018) and the socio-economic impact of cantrang ban to fishermen (Adhawati et al. 2017a, b, Hendrayana & Hartanti 2018). The results of these studies showed that: (1) there are many cantrang that has been modified by fishermen and have caused enviromental damage; (2) there are many fishermen who do not follow the requirements of cantrang provided by the government; and (3) the ban of cantrang have caused a negative social-economic impact to cantrang fishermen.

All of these studies however did not explore the alternative narratives that derive from fishermen social movement organizations. The fishermen social movement organizations, as the producer and user of cantrang, have their own definitions and operations about cantrang and have their own logic in defending cantrang as a environmental friendly fishing gear. This study wants to assert that to understand the discourse of cantrang, exploring alternative narratives of cantrang from fishermen themselves is important.

2 METHODS

In order to gather the narrative among fishermen social movement organizations, this article used a qualitative research design. In-depth interviews were conducted to members of fishermen social movement organization, such as ship crews, captains of ship, ship owners, and also leaders of social movement organization. There are total of 26 informants interviewed, where each interview lasted from 45 minutes until 1 hour. The study area were chosen purposively based on coastal area and the highest number of cantrang fishermen in Indonesia. Four areas where chosen, which are, Indramayu in West Java, Tegal and Pati in Central Java, and Lamongan in East Java. The study was parallely conducted at four districts for two weeks on July 2018.

The coding process of data includes: themes finding and using the narrative theory in dividing characterisation, plot (opening, middle, end), and moral values. Triangulation is done by conducting further interviews with informants in order to provide accurate findings. Informants were also chosen from various regions in Indonesia to gain various perspective and meanings that could reinforce the research findings. We also gather secondary data such as calendars, posters, pictures documented by the fishermen SMO's to strengthen our data.

3 THE STRUGGLE OF CANTRANG FISHERMEN

3.1 *Who is the victim? Who is the perpetrator?*

Scott and Lyman (1986) said that story always include a negotiation of identity from the characters. There are five characters identified in the struggle of cantrang fishermen. Every character contributes the dynamic of contention of cantrang ban, where violence, pressure, and anger are communicated between the characters. Those five characters are: the victim, the perpetrator, the supporter, the bystander supporters, and the opportunist.

3.1.1 *The victim: cantrang fishermen*

'We are the victim of Ibu Susi's policy…', is a common statement from cantrang fishermen regarding the cantrang ban. They thought that the cantrang ban was made without any consideration or inputs from local fishermen. The anger of fishermen brought by the sudden policy by Ibu Susi pushed them to organize a massive protest as an effort to ensure that their voices would be heard. The protest were organize by local SMOs which is a member of a national based fishermen association. There are two big national fishermen association Aliansi Nelayan Indonesia/*Alliance of Indonesia's Fishermen* (ANNI) and Himpunan Nelayan Seluruh Indonesia/*Association of Indonesia's Fishermen* (HNSI). Both of these organizations have several roles. They have to mobilize their local members to held a protest at local or national level. Fishermen from other SMOs are identified as enemy. Cantrang fishermen also have various identity such as cantrang fishermen, ship owners, captains, ship's crew, retired fishermen, activists, fish processing labor, laborers, wife of fishermen, and rickshawer, who have different interests and needs. Yet, they could merge these various identities to a collective identity: fishermen as the victim.

The effort to construct a collective identity was by using other issues, such as rise in fuel prices and changes of ship permit, to gather sympathy from non-cantrang fishermen—even labors of fisheries industries—are also the victims of government policies. They suffer from injustice, their voices are unheard, and have to fight independently for survival. They also identified themselves as people who have no interest in politics, where they rejected any kind of aids to support their activities so there are no other interests to be served in this struggle. The fishermen stated that their fight is 'a genuine struggle from fishermen', 'to fulfill our hunger (while pointing their stomach)', and 'to make my wife's kitchen busy'. One of the effort to hold fishermen interest is by organizing each member of SMOs to give contribution in funding their activities. Sometimes, ship owner are the one who responsible in providing the money, if the members can not give any contribution. These efforts showed that cantrang fishermen tried to construct a solidarity between fishermen, as a victim of hegemon and to fight the hegemon by critizing the policies.

3.1.2 *The perpetrator: Ibu Susi*

'Demote Ibu Susi from MMFA!' was the expression from one of the demonstran at the front of Presidential Palace. 'Ibu Susi' or Madam Susi is the Minister of Ministry of Maritime and Fisheries Affairs (MMFA) who is seen as a perpetrator because she is the one who issued the cantrang ban. The cantrang ban by Ibu Susi sparked the anger of fishermen when she gives a comment that '...the fishermen protests are not pure! They are getting paid by certain people! (to do the protest)" or when she stated that '...cantrang is a harmful fishing gear'. The fishermen saw Ibu Susi as a reckless person for banning cantrang without having any discussion with local cantrang fishermen, as quoted: 'we, fishermen, are patient enough, waiting for Ibu Susi to have dialogue with us. But until now, the ministry do not give any response'. The fishermen also stated: "How could she forget her roots! She used to be someone who works as fish laborers!' because she generalize one violation of cantrang use to the whole cantrang fishermen who do not operate cantrang in illegal way.

3.1.3 *The mediator: Jokowi*

In this conflictual situation, fishermen also have a supporter figure where they could express their desire. Current President Republic of Indonesia, Joko Widodo or Jokowi as people called him, is seen as a supporter in this struggle. He is seen as a mediator who can compromise with fishermen and also with Ibu Susi. Jokowi also invited fishermen representative to Presidential Palace to have a dialogue about cantrang ban where the result is the delay of cantrang ban implementation, where cantrang fishermen have enough time to change their fishing gear. This statement is seen as a win-win solution, to save the fishermen from Ibu Susi policy but also validate Ibu Susi's policy as something that should be done.

3.1.4 *The bystander supporters*

Fishermen SMOs also ask for dialogue with several government agencies, such as Legislative Assembly, Economy Industry office, President's special staffs, Ombudsman, and local fisheries and maritime offices. The government agencies act as bystander supporters, where they could facilitate a meeting so cantrang fishermen have space to express their opion about cantrang ban but do not have any power to change policy. These agencies would only send reccomendations to the policy maker.

3.1.5 *The opportunist*

Political parties are indentified as an opportunist by fishermen. Fishermen rarely showed any likeness to certain political parties, but if political parties would like to initiated feasibility studies about the use of cantrang, fishermen are willing to cooperate. Just because political parties can held a feasibility studies and give recommendation to government to reconsider the cantrang ban, fishermen realized that this move was a way for the political parties to gain followers/voters.

3.2 *Cantrang fishermen struggle: a pseudo victory?*

Table 1 shows the chronogical order of cantrang struggle. We have categorized the plots into several chapters to use what Steinmetz (1992) stated that narrative has a beginning, a middle, and an end, and toward the ends is filled with conflicts, causal explanations, and the sequence of events. We devided the narratives into four chapters: (1) The Sudden Ban; (2) Escalating Protest and Seeking Support; (3) 'Say no to Ibu Susi!'; and (4) A Pseudo Victory.

The first chapter explain the sudden cantrang ban issued by government. The fishermen responded to this ban by a series of local protest in several district in Java and escalated to a province level protest in Central Java. These protests however were not responded by the government. The second chapter describes the fishermen effort to escalate their protest. The fishermen SMOs held a national protest in Jakarta and held dialogue with several government agencies to communicate their opinion about the cantrang ban. Fishermen even sent complaint to Ombudsman, where Ombudsman gave a recommendation to policy maker to give a transition periode for cantrang fishermen to change their fishing gear. Ibu Susi granted this recommendation where she issued a Notification Letter that cantrang ban implementation will be delayed until 31st December 2016.

Ibu Susi's decision however still gave no benefit to the fishermen. This leads to chapter 3, where local fishermen started to provoke Ibu Susi by demanding to Jokowi to demote Ibu Susi from her position, rejected Ibu Susi when she visited local fishermen, even rejected the aids given by MMFA because fishermen thought Ibu Susi have make fishermen suffer enough with her policies. Several feasibility studies also showed that cantrang is not harmful to the environment and alternative solution about cantrang could be achieved, however Ibu Susi kept rejecting the result of feasibility studies. The result of feasibility studies initiated by Central Java local government stated that cantrang do not harm the environment, where cantrang did not take any mud or coral reefs and also no shrimp. While the feasibility studies initiated by political party called NasDem have showed that not all cantrang harm the environment. NasDem recommended the government not to ban cantrang, but control the production and operation of cantrang itself to ensure that the fishing gear made by fishermen do not harm the environment.

The aids from government, free boats and free gillnet, were seen as something that gives fishermen another headache. The boats went in flame only after several uses and the gillnet was seen as a fishing gear that do not produce as much as cantrang and also harm the environment. All of these attitudes showed that fishermen put all the blame to Ibu Susi. The escalating protests and provocation however did not give any benefit to the fishermen. Once the delay period given by Ibu Susi is expired, all fishermen have already plotted a massive protest to be held in Presidential Place at Jakarta and have succeeded to make President and Ibu Susi give response to fishermen demand. However, Ibu Susi's answer to this massive protest was still not lifting the cantrang ban. Ibu Susi and Jokowi kept their standpoint that cantrang have to be ban, with these conditions: (1) only Java fishermen are allowed to still use cantrang until they are ready to change their fishing gear, while non-Java fishermen have to stop using cantrang; (2) the cantrang Java fishermen could only sail in Java sea; (3) prohobition of new cantrang ship registration; and (4) cantrang ship will be arrested but instead of giving penalty, sea police have to facilite the fishermen to change their fishing gear, as quoted from Ibu Susi's statement:

Table 1. Plots of cantrang struggle by fishermen SMOs.

Chapter	Dates	Sequence of Acts
Chapter 1: The Sudden Ban	09/01/15	Government issued cantrang ban through Government Rules Fisheries and Maritime No.2/2015
	20-26/01/15	A series of protest at district level and province level regarding the cantrang ban
	26-30/01/15	Hearing with Legislative Assembly of Republic of Indonesia regarding cantrang ban
	28/01/15	Fishermen from Central Java held protests at provincial level to lift cantrang ban
Chapter 2: Escalating Protest and Seeking Support	25/02/15	National Protest by Fishermen in Jakarta, Indonesia regarding cantrang ban
	07-29/04/15	Dialogue between cantrang fishermen and governors of Central Java and President of Republic of Indonesia to lift cantrang ban
	22/05/15	Feasibility studies on cantrang, initiated by Central Java local government. Result: cantrang do not harm the environment
	25/06/15	Ombudsman reccomendation: giving transition periode minimum 2 years for cantrang fishermen to change their fishing gear
	11/02/16	Ibu Susi issued a Notification Letter where fishermen can only used cantrang until 31 December 2016, with several specifications of cantrang operation
Chapter 3: 'Say no to Ibu Susi!'	28-29/03/16	Local districts held a protest to demote Ibu Susi
	03/01/17	Directorat General of Maritime and Fisheries issued a Notification Letter on Facilitatition of Changing Fishing Gear that banned in Indonesia. Result: facilitation process for cantrang fishermen until 10 June 2017
	19/06/17	Directorat General of Maritime and Fisheries issued a Notification Letter on Extension of Transition Period for Cantrang use in Indonesia. Result: extending transition period until 31 December 2017
	08/09/17	Dialogue with Presidential Office Staffs regarding the next feasibility studies on cantrang
	24/10/17	Dialogue with Ombudsman of Republic of Indonesia and Head of Economy Industry Office regarding the socio-economic impact of cantrang ban
	1-3/11/17	Maritime expedition initiated by Political Party called Nasional Demokrat. A field survey to observe the negative impact of cantrang ban. Result: Nasdem found that cantrang ban will harm local socio-economy. Nasdem asked government to do feasibility studies that observe two things: the operation of cantrang and the fish catched by cantrang
	14/11/17	Protest of rejecting the visit of Ibu Susi in Tegal and demanded for cantrang legalization
	22-28/11/17	Cantrang feasibility studies initiated by Political Party Called Nasional Demokrat. Result: not all cantrang harm the environment. Recommendation from Nasdem to the government: there should be a standardization of cantrang, not ban of cantrang
	30/11/17-07/1201	Protest held by Lamongan (a district in East Java) fishermen: rejecting aids from MMFA, rejecting the visit of Ibu Susi, and demanding cantrang legalization
	08/01/18	Protest held by fishermen from Java Island, Sumatera Island, and Madura Island
	15/01/18	Dialogue between fishermen representatives and President in Tegal: demanding cantrang fishermen to sail again
Chapter 4: A Pseudo Victory	17/01/18	Massive protest of Indonesia's fishermen in front of Presidential Palace to lift the cantrang ban nationally.
	18/01/18	Ibu Susi won't lift Cantrang ban, fishermen who registered new cantrang boat will be arrested by sea police and will be facilitated to change their cantrang gear.

Source: Calendar 2018 of ANNI's cantrang struggle.

Cantrang ban is delayed only in 6 area: Batang, Tegal, Rembang, Pati, Juwana, and Lamongan. Other than those, there is no excuse. During this transition period, cantrang fishermen can sail but only in Java Sea and do not register a new cantrang ship.

(CNN Indonesia 18th January 2018)

The current condition of cantrang struggle still left a bad taste for cantrang fishermen. The struggle of cantrang fishermen were filled with anger, conflict, and provocation. While they are succesful in organizing a massive protest and could bring Jokowi and Ibu Susi to respond to their voices, the result did not meet cantrang fishermen demand. Cantrang fishermen still have to change their fishing gear, even though they believe that cantrang do not harm the environment.

3.3 *Cantrang do not harm the environment—regardless of what government said*

There are at least two level of stories in social movements: personal and collective level stories, although these two levels do not entirely separate nor it run together (Davis 2002). Personal level stories or participant narratives are the stories people tell about themselves—their own experience, such as significant life moments or traumatic events, while collective level stories or movement narratives are the stories that represent the purposes and the paths of actions of the movement (Benford 2002). This section explains the narratives from both level.

3.3.1 *Participant narratives: the retired fishermen*

A retired fishermen from Juwana, Pati who has been sailing almost all of his life explained the differences between cantrang, trawl and gillnet. He had tried varities of fishing gear and also made varities of fishing gear, including cantrang, trawl and gillnet. Other members of social movement organizations stated that 'we believe everything he said about fishing. He is a respected figure for us'. Interview conducted with other informants about cantrang, trawl, and gillnet also have similarity with the narratives of retired fishermen. He also had invited by Indonesia's president to explain about cantrang, he even stated in the interview:

"...I've already spoke about this a thousand times even to the President. Sometimes I feel tired to explain (about cantrang) to people who came to me. But I feel like I have to tell you the truth".

(Interview, 23rd July 2018).

The retired fishermen stated that '...cantrang is a traditional fishing gear. It has been inherited to us by our ancestors. The ban will definitely harm our life'. He also said that government have always sees cantrang as trawl, while according to him cantrang and trawl is really different. Cantrang is a fishing gear that operates by lowering (using a weighted bed) the net in a straight line (while boat is stand still) and cantrang do not pull the net while moving the boat. Cantrang also doesn't have an opening mouth, so the net is scooped rather then pulled. To catch fish, cantrang fishermen have to scoop the net approximately 6-10 times in one or different places. It only takes an hour for cantrang to fishermen to do a scoop. Cantrang also do not operates until ocean floor, and won't harm the coral reefs. 'Our nets will be damage if we accidentally catch coral reefs', stated the retired fishermen to ensure that fishermen are most likely incur losses if they purposely harm the coral reefs. Government also said that the mesh size of net is below 2 inches so small fishes could also be catch. However, the retired fishermen, stated that the cantrang he made is above 5 inches, so it didn't catch any small fishes. He clearly stated that cantrang do not harm the environment.

Trawl, however, that operates by lowering the net until the ocean floor and pull the net while the boat is moving. Trawl also have an opening mouth, so while the boat is pulling the net the mouth will always open to catch fish. To pull trawl, fishermen spend a long time until they get a lot of fish. The mesh size of trawl net is 3 inches, so it definitely catch small fishes. According to the retired fishermen, trawl definitely harm the environment. The quotation below explains his understanding about trawl and cantrang:

...trawl has water board. I was a trawl fishermen for 7 years. The net will always open as long as fishermen wishes. Cantrang is different. It is not pulled because it doesn't have water board. Pulling the net will only waste the time and fuel.

(Interview, 23rd July 2018)

All of his narratives are gathered from his experience becoming a fishermen for almost 45 years who have tried various fishing gear. Gillnet, an alternative fishing gear of cantrang provided by government, on the other hand also harm the environment, according to the retired fishermen. He stated that '...gillnet sweep the ocean floor and catch protected fish'. Gillnet, as an alternative fishing gear provided by government, is more damaging than cantrang, according the retired fishermen.

The narratives shared by the retired fishermen is actually a bundle of claims that constructed because of the sequence of acts with government. The main reason from government to ban cantrang is because of its environmental damage, thus the retired fishermen tried to construct his narrative, in a detailed way, to explain the production and operation of cantrang. He even compare cantrang to trawl because government still considered cantrang as a modified trawl. After constructing narrative about cantrang and trawl, he also explain narratives about gillnet that later become an alternative fishing gear promoted by the government. CMM theory supports our analysis that the narratives are being co-constructed through sequence of acts. Every communication events between fishermen and government leads to another motivation to counter narratives.

3.3.2 *Movement narratives*

The success of fishermen for gathering masses and held a massive protest was possible because of the organizations to mobilize local and national organization. The organization could provide human resources or activists that linked the fishermen to other organizations who have the same interest to spread the narrative of cantrang fishermen. Benford (2002) assert that social movement actors seek to sustain the narrative shared by the members. Movement narratives or narratives that shared by the members of SMOs are bundles of personal stories that have been sustained by the members as the true interest of SMOs. Benford (2002) also stated that movement narratives refer to the various myths, legends, or folk tales that collectively constructed by the participants about the movement and things that they would like to change. Benford named this effort as a 'social control within social movement' to control what kind of narratives that should be sustained in SMOs.

The social control conducted by the fishermen SMOs are: (1) linked fishermen with related government agency so that they could express their opinion or demand, we could see from the plot explained in Table 1 that the SMOs could arrange meetings between legislative assembly, ombudsman, etc.; (2) let the fishermen speak freely according to their experiences, however the SMOs sometimes encourage a figure that respected by the members of organizations to tell his stories, so this stories could be heard by the members and become a shared narrative. The retired fishermen is one of the figure respected by the members, where his narratives is seen as the truth of their movements; and (3) producing organizations artifacts to communicate their identity or interest in cantrang struggle. Figure 1 showed as the design of banner as an artifact at the protest where this design is provided by the national organizations. The national organizations sent these to local organizations staffs to be printed or to be shared to other members so each local organizations could bring the banners at the protest.

As shown in Figure 1, statement of fishermen social movement organizations are (from top-left to bottom-right): 'give us our rights. Legalise cantrang', 'proven: cantrang is environment friendly', 'cantrang is the creativity of Indonesia's fishermen', 'Honorable Mr. President help us, we need to eat', 'Honorable Mr. President help us, our children need to go to school', 'Honorable Mr. President, help us we are being victimized by MMFA', 'KKP(MMFA) = Kok Kami jadi Penggangguran (we became jobless)', 'Susi is not viable to becoma a minister of MMFA', 'demote Susi! Until when we suffer from this', 'fishermen are also human, we need to eat, our children need to go to school, legalise cantrang', 'Honorable Mr. President, choose! Susi or fishermen?', 'why do cantrang fishermen are forbidden from thinking!!!!', 'we

Figure 1. Design of posters for cantrang protests.

are fishermen, not robbers!' and 'do not kill fishermen! We are not criminals!'. All of these statement reflects the narratives of the movement. The fishermen demand the legalisation of cantrang because it doesn't harm the environment, they put the blame on Ibu Susi, the cantrang ban will give negative impact to their life, and other provocation that emphasizing the needs to legalise cantrang in Indonesia.

3.4 *Moral value of the story*

3.4.1 *Fishermen vs Government, do they communicate effectively?*

The sudden ban of cantrang by government raised several questions from the fishermen: why did the government suddenly ban cantrang? why did the government do not have a dialogue with cantrang fishermen to discuss about the operation of cantrang? and are banning the cantrang is the only solution? These questions express the relationship between the people and the government: are Indonesia's government democrative enough so that any policy changes are conducted through a participatory approach where the voice the people should be heard? Is it ethical to ban cantrang without having any discussion with fishermen? The plof of cantrang struggle showed us that fishermen have the spaces to communicate their opinion, however the decision has been made. Government chose to compromise rather than to reconsider the cantrang ban policy.

Jokowi, in this narratives, have been more lenient to the fishermen. He granted the wish of fishermen to have a dialogue with him about cantrang and he also tried to help the cantrang fishermen by delaying the cantrang ban. Yet, Jokowi couldn't change Ibu Susi's decision since it also based on the government findings about the harm of cantrang to the environment. Jokowi's attitude as a leader showed that he is a parent figure where he should protect the feeling of his childrens. Both Ibu Susi and fishermen's feelings have to be protected by listening to both of them. The leadership of Jokowi is influenced by his Javanese roots, where a leader should be like a father, protecting people, tolerant, and have moral integrity (Irawanto et al. 2012, Selvarajah et al. 2016).

3.4.2 *Cantrang, does or does not harm the environment?*

The main issue of cantrang, based on government finding, is its negative impact to the environment. However, does diminishing cantrang is the best solution in protecting the environment? Or are enviromental issues have to be prioritized over people's life? In environmental ethics, the shallow ecology movement is defined as the fight against pollution and resource depletion where the central objective of which is the health and affluence of people in the developed countries (Næss 1973). It is clear that Ibu Susi leans to shallow ecology movement where conservations is the only answer of resource depletion. However, fishermen have construct alternative narratives where the use of cantrang in each districts have differences in operating it. The retired fishermen also argued that there a lot of fishermen who have modified their fishing gear but still called it cantrang even though it doesn't work like cantrang. The alternative narratives given by the retired fishermen suggests that an objective and detailed observation about cantrang is needed.

3.4.3 *Do protest and dialogue result to win-win solution?*

The plots of cantrang struggle also raised a question about the function of democratic government and people's way in communicating with rulers. On one side, government was reckless for banning cantrang without giving any chances to cantrang fishermen to speak for themselves and without thinking any alternatives that does not risk cantrang fishermen's life. Do these moves reflect a democratic government? The compromises at the end of cantrang struggle also do not give any benefit to cantrang fishermen as they have to change their fishing gear and their alternative narratives are basically goes unheard because government do not reconsider it. Provocation and protests by fishermen also gave no power to the fishermen, except the massive reporting of the protest and dialogue occured with the policy makers. This showed us that provocation and protests do not granted the wishes of fishermen.

4 CONCLUSIONS

The struggle of cantrang fishermen showed us that narrative is dynamic, where the story is being co-constructed by sequence of acts between fishermen and government. The main issue of cantrang struggle is cantrang's negative impact to the environment, however the story leads to other things such as: the operation of cantrang, differences of cantrang operation in each location, blaming Ibu Susi, the negative socio-economic impact of cantrang ban, and feasibility studies that goes unheard. As CMM explained, the sequence of acts also defined the relationship between fishermen and government, when government suddenly banned cantrang, the reaction of cantrang fishermen is protesting since they felt they are left behind on something dear to them. When Ibu Susi respond to the fishermen's protest as a paid protest, the fishermen retaliate by putting all the blame on Ibu Susi, rather than explaining their stand on cantrang ban. This showed the fishermen's frustation since Ibu Susi ignored cantrang fishermen voices. The sequence of acts did not give any spaces for both sides to discuss about the issues, but it become worse by provocation from both sides.

The narratives also showed moral elements. The narratives emphasized the relationship between fishermen and government to achieve a transformation: an environmental transformation by government and a socio-economic transformation by fishermen, the environmental ethics of cantrang, and the communication ethics in achieveing social change. The moral elements of the narratives also give an pratical implication from this study: that giving spaces for the people to communicate is not enough, government also needs to really listen to people's voices. Government have to listen and to be more open to alternative narratives, to ensure that a sound policy is made. A good communication skills of government figure is also needed, as it is showed that Ibu Susi's statements about fishermen protests provoked the anger of fishermen.

REFERENCES

Adhawati, S.S., Baso, A., Malawa, A. & Arief, A. 2017a. Social study of cantrang (danish trawl) fisheries post moratorium at Makassar Straits and Bone Gulf, South Sulawesi Province, Indonesia. *AACL Bioflux* 10(5): 1140-1149.

Adhawati, S.S., Baso, A., Malawa, A. & Arief, A. 2017b. Comparative study of economic value post cantrang moratorium on the waters of the Gulf of Bone and Makassar Straits, South Sulawesi Province. *International Journal of Oceans and Oceanography* 11(2):201-215.

Benford, R.D. & Snow, D.A. 2000. Framing process and social movement: an overview and assessment. *Annual Review of Sociology* 26(1): 611-639.
Benford, R.D. 2002. Controlling narratives and narratives as control within social movements. In Davis, J.E. (ed.). *Stories of Change: Narrative and Social Movement*. Albany: SUNY Press.
Boudana, S. & Segev, E. 2017a. The bias of provocation narratives in international news. *The International Journal of Press/Politics* 22(3):314-332.
Boudana, S. & Segev, E. 2017b. Theorizing provocation narratives as communication strategies. *Communication Theory* 27(4):329-346.
CNN Indonesia. 2018. 'Menteri Susi Izinkan Cantrang Hanya di Enam Wilayah'. From https://www.cnnindonesia.com/nasional/20180118103318-20-269811/menteri-susi-izinkan-cantrang-hanya-di-enam-wilayah accessed on 18th December 2018.
Davis, J.E. 2002. Narrative and social movements: the power of stories. In Davis, J.E. (ed.). *Stories of Change: Narrative and Social Movement*. Albany: SUNY Press.
Driessens, O. 2013. 'Do (not) go to vote!' Media provocation explained. *European Journal of Communication* 28(5):556-569.
Fraser, N. 1992. Rethinking the public sphere: A contribution to the critique of actually existing democracy. In C. Calhoun (ed). *Habermas and the public sphere*. Cambridge: MIT Press.
Gumucio-Dagron, A. 2009. Playing with fire: power, participation, and communication for development. *Developmen in Practice* 19(4–5):453–465
Hendrayana & Hartati, N.U. 2018. The impact of cantrang ban on socio-economic conditions of Tegal fisheries. *In National Proceeding of National Seminar Edusainstek*. Semarang: Universitas Muhammadiyah Semarang.
Hendroyono. 2018. Socio-economics of trawl fisheries in Southeast Asia and Papua New Guinea. In Slar, S.V., Suuronen, P. & Gregory, R (eds.), *FAO Fisheries and Aquaculture Proceeding; Proceeding of the Regional Workshop on Trawl Fisheries Socio-economics, Vietnam, 26-27 October 2015*. Da Nang: FAO.
Herman, D., Phelan, J., Rabinowits, P.J., Richardson, B. & Warhol, R. 2012. Narrative theory: Core concepts and critical debates. Ohio: Ohio State University Press.
Houghton, J., Siegel, M. & Goldsmith, D. 2013. Modeling the influence of narratives on collective behaviour case study: using social media to predict the outbreak of violence in the 2011 London Riots. *Working Paper CISl# 2013-15*. Cambridge: Massachusetts Institute of Technology.
Irawanto, D.W., Ramsey, P.L., & Tweed, D.C. 2012. Exploring paternalistic leadership and its application to the Indonesian public sector. *International Journal of Leadership in Public Services* 8(1):4-20.
Jennifer, E. 2015. The Future of Social Movement Organizations: The Waning Dominance of SMOs Online. *American Behavioral Scientist* 59(1):35–52
Liu, C.W. 2016. Student activism, institutional amnesia, and narrative (re)construction: lessons from Brendeis University's #FordHall2015 protests. *eJournal of Public Affairs* 5(2): 234-249.
Næss, A. 1973. The shallow and the deep, long-range ecology movement. *Inquiry* 16:151-155.
Pearce, W.B. & Pearce, K.A. 2000. Extending the theory of Coordinated Management Meaning (CMM) through a community dialogue process. *Communication Theory* 10(4):405-423.
Polletta, F. 2006. *It was like a fever: storytelling in protest and politics*. Chicago: University of Chicago press.
Polletta, F. & Chen, P.C.B. 2012. Narrative and Social Movement. In Alexander, J.C., Jacobs, R.N. & Smith, P. (eds.). *The Oxford Handbook of Cultural Sociology*. Oxford: Oxford University Press.
Sasmita, S., Martasuganda, S. & Purbayanto, A. 2012. Technical design of danish seine on North Java Waters. *Jurnal Perikanan dan Kelautan* 2(2):79-86.
Selvarajah, C., Meyer, D., Roostika, R. & Sukunesan, S. 2016. Exploring managerial leadership in Javanese (Indonesia) organisations: engaging Asta Brata, the eight principles of Javanese statesmanship. *Asia Pacific Business Review* 23(3):373-395.
Scott, M.B. & Lyman, S.M. 1986. Accounts. *American Sociological Review* 33(1):46-62.
Steinmetz, G. 1992. Reflections on the role of social narratives in working-class formation: narrative theory in social sciences. *Social Science History* 16(3):489-516.
Wahyuningrat, Haryanto, T. & Rosyadi, S. 2018. Practices of Illegal Fishing in Pemalang Region: a policy analysis. *E3S Web of Conferences* 47:1-7.

Rural Socio-Economic Transformation – Kinseng et al. (Eds)
© 2019 Taylor & Francis Group, London, ISBN 978-0-367-23603-8

Social engineering of local government in the perspective of communication study toward the development program of pedestrian area

T. Yuniarti
Faculty of English and Communication, Islamic 45 University, Bekasi, West Java, Indonesia

A. Saleh, M. Hubeis & R.A. Kinseng
Department of Communication and Community Development Sciences, Faculty of Human Ecology, Bogor Agricultural University, Bogor, West Java, Indonesia

ABSTRACT: The development of pedestrian areas in Bekasi to be urgent because pedestrians tend to be more vulnerable to traffic accidents and need safe spaces from cars or motorcycles. Hence, the present study aims to analyze social engineering of Bekasi government toward the development of pedestrian areas in the perspective of communication study. The present study applies a descriptive qualitative method to obtain data by observing and interviewing key informants. The results suggest a cooperation of local government and various stakeholders in promoting the program as well as maintaining facilities of pedestrian areas. The co-operation is conducted to promote the responsible use of pedestrian facilities among pedestrians and other road users. An effective communication to public can be performed better via social media because information related to government' programs or events can reach public faster and more accessible.

1 INTRODUCTION

Traffic jams might lead to traffic accidents as well as the increase of death rate among pedestrians, making pedestrian areas important as safe spaces for pedestrians. Tamin (1992) states that traffic jam usually takes place in cities with population more than two millions. This problem also happens in Bekasi, one of cities in West Java. Located near the border of Eastern part of Special Capital Region of Jakarta (DKI Jakarta), Bekasi has a population of 2.8 million of people. According to the medium-term development (RPJMD) of Bekasi, volume ratio of road capacity of Bekasi is 0.83 percent, indicating congested road conditions (Pemkot Bekasi 2016) . This condition might lead to the increase in death rate due to traffic accidents. In 2013, World Health Organization stated that the pedestrians' death rate could reach 22% or even 67% (in some countries) of total death rate due to traffic accidents (WHO 2013). It is why building pedestrian areas becomes an important program for Bekasi government to avoid the numbers of traffic accidents.

According to Urban Safety Management (USM), developing pedestrian areas is an element that should be fulfilled to provide public safety in every city. Quimby et al. (2003) argues that USM is an integrated and systemic approach which prevents and reduces numbers of traffic accident by involving various approaches of different studies, point of views, and stakeholders. It is because USM involves a careful planning in which views and opinions from different parties (civilians, practitioners, politicians/policy makers) are considered. In addition, USM has been proven successful to reduce death rates due to traffic accidents in several countries (Quimby et al. 2003). It can be said that USM is a suitable approach to develop public facilities, especially pedestrian areas.

Besides USM approach, good communication is a key success to socialize a government's program. In this regard, communication can be defined as a relational process to create and interpret responsive messages (Griffin 2012). As one of the most congested cities in Indonesia, building pedestrian facilities becomes a main program of Bekasi government. To ensure the success of program, Bekasi government needs to socialize the development and facilities' maintenance to public so that the program's benefits can reach public.

In this sense, social engineering can be one of solutions. Social engineering refers to an approach to compromise with a group of individuals through communication strategies as well as influencing through political, economy, or social power (Everett 2006, Krombholz et al. 2014). Several studies have investigated the effects of social engineering through the perspectives of communication studies. Madanijah et al. (2004) used social engineering to conduct an experiment on media counseling on knowledge and attitudes toward staple food diversification in Bogor and its regency area. On the other hand, Gasic (2013) investigated the accessibilities of information in pedestrian areas for disable people. In addition, Hoekstra and Weigman (2010) conducted a study on the effectiveness of road safety campaigns while Nathanail and Adamos (2012) investigated efficient strategies for road safety campaigns.

A study of Everett (2006) shows that knowledge or information in social engineering can be a reference to local stakeholders such as civil society leaders entrepreneurs, citizens and public officials at every level to build local control and accountability. Furthermore, Krombholz et al. (2014) elaborated social engineering as an art that allows its users to compromise with information system. In this sense, information access can be used to influence and persuade. In addition, Nathanail & Adamos (2013) state that communication can be used as an efficient strategy to approach public. Besides that, communication can also be used to relay messages of pedestrian facilities to disabled pedestrian users (Gacic 2013). Thus, pedestrian facilities should be built by considering human rights. Information regarding pedestrian facilities should be accessed by everyone, including disable people. Therefore, every facility in pedestrian facility should be completed with sufficient information for every pedestrian user.

In regards of communication, social engineering can be interpreted as a way to influence public through a systemic and planned message delivery. Rogers (1983) said that communication is a process where an individual creates and shares information to other people in order to reach a similar understanding. Madanijah et al. (2004) argues that a message delivery through a combination of conventional and electronic mass media can result in big influences toward the improvement of knowledge and attitudes of food products' diversification of local government of Bogor and its regency. In order to improve the communication, a study of Hofmann et al. (2013) toward external communication of 25 cities in Germany recommend an improvement in online communication that should be conducted by local governments to communicate with public. Furthermore, a study of Lovari & Parisi (2011) on the use of Facebook' page in four regencies in Italy, suggests a content combination that allows a communication flow from local government to public.

A study which investigates social engineering of local government, especially Bekasi from the perspective of communication study has not been performed yet. This study seems beneficial to provide stakeholders with a better social engineering concept. Therefore, the present research aims to analyse social engineering conducted by Bekasi government to communicate the development and maintenance program of pedestrian areas to public. In the present study, the analysis is conducted to provide a description of effective message deliveries. This analysis will be useful to either local or central government to provide better communication programs for similar future projects.

2 METHODS

A descriptive qualitative method is applied to obtain data. Furthermore, a description on social engineering of Bekasi government to build pedestrian areas will be analyzed in the perspective of communication study.

USM's (Urban Safety Management) approach has been developed in Netherland and England. In addition, the USM approach pinpoints the development of pedestrian areas as one of elements that needs to be realized to provide public safety (Quimby et al. 2003). The informants of the present study are stakeholders of Bekasi government as well as public. Informants from Bekasi government are chosen because they have obligations to deliver government's programs to public. The total of informants from Bekasi government is 12 people. These people are structural officials of Bekasi government who are involved in the present study: Rahmat Effendi (Mayor of Bekasi), Tri Adhianto (Deputy Mayor of Bekasi), Titi Masrifahati (Head of Office of Communication, Information, Coding and Statistics/Diskominfostandi), Widayat Subroto Hadi (Head of Highways Agency of Office of Public Works and Housing/PUPR of Bekasi), Sayekti (Head of Public Relations of Bekasi), and Gutus Hermawan (Head of Bekasi Timur sub-district). On the other hand, a group from public was chosen purposively because they intersect with local government' programs, especially those who are involved with pedestrian areas. They are Alfred Sitorus (Head of Pedestrians' Coalitions), Sutanto (Head of Communication Forum of Hamlets of Bekasi Timur, an anonymous street vendor, and an anonymous taxi bike rider.

Data is obtained by interviewing key informants such as Rahmat Effendi (The Major of Bekasi), Widayat Subroto Hardi (Head of High-ways Agency of Office of Public Works and Housing (PUPR) of Bekasi), and Titi Masrifahati (Head of Communication, Information, Statistics and Coding Services (Kominfostandi) of Bekasi). In addition, an observation to a pedestrian area in Jalan Ahmad Yani (one of main roads in Bekasi) as well as to the documentation of development and communication processes of Bekasi government to public will be analyzed. Data triangulation will be conducted by integrating and categorizing data which obtained from interviews, field observation, documentation, and literature review.

Documentation of the present study was obtained from communication events conducted by Bekasi government to pedestrians, street vendors, and taxi bike riders regarding pedestrian areas. Observation method of present study is a non-participant observation in which data is obtained without getting involved into the situation. Observation was performed to communication activities of Bekasi government to people living in pedestrian areas, mass media, and social media.

3 RESULT AND DISCUSSION

3.1 *The development of pedestrian area in Bekasi*

The development of pedestrian areas in Bekasi is ruled in Regional Regulation No. 5 of 2016 on Spatial Detail Plan (RDTR) of Bekasi from 2015 to 2035 Article 59 which states that the development planning of pedestrian areas of Bekasi consists of a space for pedestrians on the road side; on the side of building; on the side which borders with a water body or river; above ground (overpass) and underground (tunnel). In this regard, pedestrians refer to people who walk in certain distance whereas a path that accommodates the activity is called a pedestrian way (Cutler in Kusbiantoro et al. 2015). In 2018, Ministry of Public Works and Housing (PUPR) released a guide for developing pedestrian areas, especially on technical planning for pedestrian facilities. In addition, as stated in Act No. 22 of 2009 Article 25 about traffic and land transportation states that each road used for public traffic needs to be equipped with facilities for pedestrians and disable people. The afore-mentioned regulations suggest the importance of pedestrians' facilities.

To suit the needs of pedestrians, pedestrian's facilities should be built based on that shown in Figure 1.

Figure 1 shows the development of pedestrian area which consists of building boundary rules, vehicle lanes, facility lanes, and pedestrian paths on the front of building. The bicycle lane is located on the right side of sidewalk or the left side of pedestrians. Although a bicycle lane is given 1.25 m of space, it should still consider a minimal space of 1.5 m for pedestrians. Following figure shows the perspective and dimension of a bicycle lane on the sidewalk.

Figure 2 describes rules and size to build lanes for vehicle, bicycle and pedestrians. The descriptions can be a reference for local government when building pedestrian lanes.

According to Hardi as a key informant, the development of five to seven km of pedestrian areas was conducted in two years from 2016 to 2017. Besides reaching the target of program,

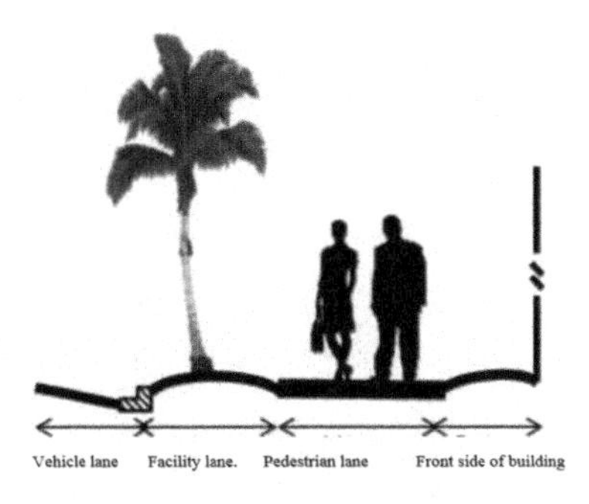

Figure 1. The example of zone division for pedestrian areas (Ministry of Public Works and Housing 2018).

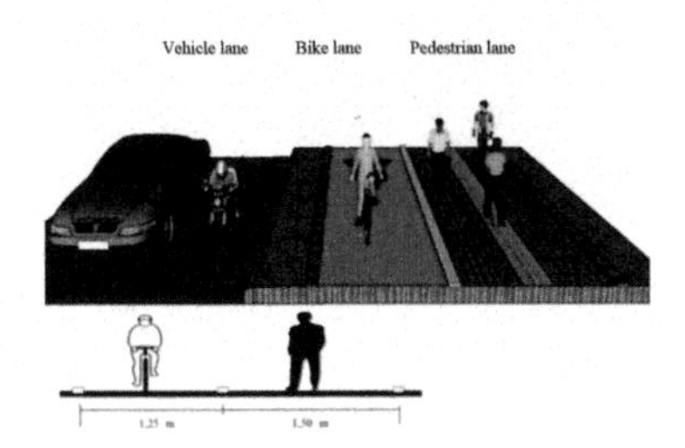

Figure 2. Perspective and dimension of shared paths (Ministry of Public Works and Housing 2018).

pedestrian areas were built in Jalan Ahmad Yani, one of the most congested roads in Bekasi. Such development was also performed in Jalan KH Noer Ali Kalimalang, Jalan Hasibuan, Jalan Chairil Anwar, and Jalan Pekayon-Pondok Gede which serve as main roads of Bekasi. Hardi said that the development has surpassed the target due to public's needs such as socialization or car free day. In addition, Hardi said that the development would be continued in 2019 as well as some road maintenance by providing utility facilities underneath the sidewalk along the pedestrian areas (ducting utility). This utility is used to control electrical and other celluler providers' cables in order to make Bekasi looks neater.

Effendi as the Major of Bekasi stated similar comments toward the ducting utility. In addition to public facilities, these pedestrian areas will serve as a recreation center during car free day events or else. Effendi also states that in 2019, the focus will be on the aesthetic side of pedestrian areas and how to accommodate 2.7 Million of Bekasi people. Following is the map of the development of pedestrian areas.

Figure 3 shows main roads become the priorities for the development of pedestrian areas. There are educational centers, business centers, train station, bus station located on those roads, making them the representation of Bekasi. The concept of pedestrian areas' development in Bekasi consists of building lines on both sides of street, 3 m width of dimension of pedestrian' lane (it can accommodate 3-4 people by assuming 1 person= 0.8 m and 2 people= 1.5 m), adding bench, adding trees for shading and orientation of pedestrians. Below is the development plan and existing pedestrian areas in Jalan Ahmad Yani.

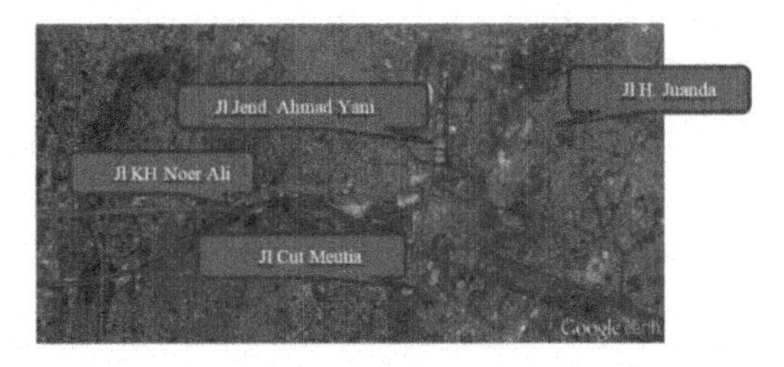

Figure 3. The development plan of early stages of pedestrian areas (Journal of City Planning of Bekasi Government 2015).

Figure 4. The Development Plan of Pedestrian (Left) (Public Relations of Bekasi Government 2018) and Pedestrian Facilities in Jalan Ahmad Yani (Right) (Author's documentation 2018).

Figure 4 shows a similarity between the development plans of pedestrian facilities in the main street and the newly built pedestrian areas. The newly-built pedestrian facilities consist of a pedestrian path, a bicycle lane, park, bench, trash can, street light, and road barriers. During car free day events, pedestrian areas are used by public to exercise or do other activities. As reported by *beritasatu.com* (2017), the development of pedestrian areas adopts the urban design concept which has been implemented in other countries. This concept involves building a road which is equipped with other facilities such as building pedestrian areas, bicycle lanes, park, bench, and street lights. Besides that, the road is also equipped with Wi-Fi and *Closed Circuit Television* (CCTV). During car free day events, pedestrian areas are used by public to exercise or do other activities. To build the pedestrian areas in Bekasi, Hardi said that the local government has spent 90 billion Rupiah. Besides the existing pedestrian areas, Bekasi government has also planned to build pedestrian areas near the rivers or water front to realize the concept of waterfront city by maximizing potentials along the rivers in Bekasi.

3.2 *Social engineering in the perspective of communication study*

Sanders and Canel (2012) argue that government's communication adopts two complementary epistemological strategy approaches; apriori and posteriori. The former depends on its analysis of variety of communication characteristics. On the other hand, a posteriori approach is conducted by analyzing empirical research outside of the organization. The former depends on its analysis of variety of communication characteristics. On the other hand, a posteriori approach is conducted by analysing empirical research outside of the organization. Government's communication can be understood as a way to identify strengths and weaknesses of its political communication contribution as well as a system that connects government and public.

In the present study, Bekasi government performed several methods to deliver information related to the development of pedestrian areas. Following are communication strategies of Bekasi government:

1) Face to face communication

This type of communication is directly performed to public during various events or observation schedule of running development. According to Effendi, caution and warning are given to public through government offices which are appointed to maintain existing pedestrian facilities. Besides that, they are also given during public events such as car free day in Jalan Ahmad Yani.

Effendi said that government should encourage positive cultures such as orderliness and discipline in maintaining cleanliness of public areas. It is why he said that ongoing socialization to public is important to maintain the existing public facilities. Such messages are given by Effendi during the face to face communication events. Following picture shows how Effendi gave instructions about the development of pedestrian areas of Bekasi to his staffs.

Figure 5. The Major Bekasi when he observed the pedestrian' construction (Poskotanews.com 2016).

Effendi said that local government delivers messages to public by collaborating with local government agencies so that they can support each other in delivering messages to public. Furthermore he elaborated that Kominfostandi acts as a coordinator in delivering messages and information whereas Public Relations of Bekasi delivers messages to public through press/journalist.

2) Mass media communication

Besides the Major of Bekasi, the Public Works and Housing agency of Bekasi (PUPR) as an executor of the pedestrian development of Bekasi also conducts face-to-face communication with public through journalists when they inaugurate pedestrian areas and a public park in Jalan Rawa Tembaga. Figure 6 below shows a socialization program of PUPR of Bekasi to a group of journalists. PUPR of Bekasi explained their plans to build rivers with the concept of waterfront city by integrating pedestrian infrastructures with its facilities.

This event was conducted on 26 December 2017 by PUPR of Bekasi to deliver their plans related to the development of pedestrian areas in Bekasi. The results of this socialization program are news in mass media which would be consumed by public.

Figure 6. The meeting of PUPR of Bekasi with journalists (Author's documentation 2017).

Figure 7. Information on the construction of pedestrian parks through social media (@dinas_pupr 2018)

3) Social media communication

Bekasi government also uses social media in communicating the messages of development of pedestrian development to the community. According to the Head of the Office of Communication, Information, Statistics and Coding (Diskominfostandi) of the City of Bekasi, Masrifahati, every service within Bekasi government is required to have a social media account as a means of communicating to the public. Accounts on social media owned include Twitter with @pemkotbekasi account (Bekasi City Government), @dinas_pupr (Bekasi PUPR Office), @humasbekasikota (Bekasi Government Public Relations). Figure 7 presents an example of a pedestrian development message delivered via the Public Works Office of Bekasi City twitter.

Based on Figure 7, the message of pedestrian development delivered through social media turned out to be quite effective in getting a response from the community, evidenced by the number of comments, retweeting (republishing the uploaded message on the twitter account), and the response that expressed likes.

According to the communication method performed by Bekasi government, communication endeavors through social media is perceived as a more effective way to obtain information from Bekasi government. It is because information from social media is received faster by public. According to Susanto as a Head of Communication Forum of Hamlet and Neighborhood Association of Bekasi, the existence of WhatsApp and other social media seem to be more effective ways to deliver information to public. The opinion is shared by Gutus Hermawan (Head of Bekasi Timur sub-district) who said that social media is a faster way to reach public. Invitations to public events are often sent via social media because they are faster and more effective. According to Tri Adhianto (Deputy Mayor of Bekasi), every neighborhood association and hamlet of Bekasi, heads of sub district, heads of department, as well as heads of district need to make a Whatsapp group to promote government' programs faster.

4 CONCLUSION

Communication with the community is not only effective through face-to-face, but needs to be through other channels so that the message reaches the community. This is the background of the importance of doing social engineering in a communication perspective. In this case the Bekasi City Government uses mass media and social media to communicate the messages of pedestrian development to the community. Information published through mass media tends to be more trusted by the public compared to information delivered directly by the Government through its media. While communication on social media accelerates the interaction between Government and society. Information on the construction of pedestrian facilities delivered through various channels is an effective way of disseminating plans and developments in the means of development. An effective communication to public can be performed better via social media because information related to government' programs or events can reach public faster and more accessible.

REFERENCES

Everett, M.A. 2006. *From social engineering to social movement: Power sharing in community change in New York's Hudson Valley and Catskill Mountains* [Dissertation]. Rotterdam (NL): Erasmus University of Rotterdam.

Gacic, A., 2013. Treatment of pedestrian communication accessibility in the process of urban design: Case study of institute for orthopedic surgery "banjica". *Facta Universitatis Series: Architecture and Civil Engineering* 11(1): 61-70.

Griffin, E.A. 2012. *A First Look at Communication Theory*. Eight Edition. New York: McGraw Hill.

Hofmann, S., Beverungen, D., Räckers, M., & Becker, J. 2013. What makes local governments: Online communications successful? Insights from a multi-method analysis of facebook. *Government Information Quarterly* 30(4): 387–396.

Hoekstra, T. & Wegman, F., 2011. Improving the effectiveness of road safety campaigns : Current and new practices. *IATSS Research* 34(2):80–86.

Krombholz, K., Hobel, H., Huber, M., Weippl, E. 2014. Advanced social engineering attacks. *Information Security and Applications* 22:13–122.

Kusbiantoro, B. S., Natalivan, P., Aquarita, D. 2015. Kebutuhan dan peluang pengembangan fasilitas pedestrian pada sistem jalan di perkotaan. *Tata Kota* 3: 65-78.

Lovari, A & Parisi, L. 2011. Public administrations and citizens 2.0: Exploring digital public communication strategies and civic interaction within Italian municipality pages on facebook. In Comunello F, editor. *Networked Sociability and Individualism: Technology for Personal and Professional Relationships*. Pennsylvania (AS): IGI Global. 238-263.

Madanijah, S., Khomsan, A., Martianto, D., Djamaluddin, M. D., Briawan, D. 2004. Kajian rekayasa so-sial dan teknik edukasi dalam diversifikasi konsumsi pangan pokok. Seminar diseminasi hasil-hasil penelitian Departemen Gizi Masyarakat dan Sumberdaya Keluarga Tahun 2003, Bogor, 6 July 2004. Bogor: IPB.

Ministry of Public Works and Housing. 2018. Pedoman Bahan Konstruksi Bangun dan Rekayasa Sipil: Perencanaan Teknis Fasilitas Pejalan Kaki. Jakarta: PUPR.

Nathanail, E. & Adamos, G., 2013. Road safety communication campaigns : Research designs and behavioral modeling. Transportation Research Part F: Psychology and Behaviour, 18, pp.107–122. Available at: http://dx.doi.org/10.1016/j.trf.2012.12.003.

Niman, M. 2017. Dipercantik, Kota Bekasi bangun jalur pedestrian. Beritasatu.com. Rubrik Aktualitas. Beritasatu [Internet].[download 2017 Jan 6]. Available at: http://www.beritasatu.com/aktualitas/412664-dipercantik-kota-bekasi-bangun-jalur-pedestrian.htmlestrian.

[Pemkot] Pemerintah Kota Bekasi. 2016. Peraturan Daerah Kota Bekasi Nomor 1 Tahun 2016 tentang Perubahan Atas Peraturan Daerah Kota Bekasi Nomor 11Tahun 2013, tentang Rencana Pembangunan Jangka Menengah Kota Bekasi. Bekasi: Pemerintah Kota Bekasi.

Quimby, A., Kirk, S.J., & Fletcher, J. 2005. Urban safety management: Guidelines for developing countries annexe 2 applying the urban safety management approach in Cirebon, Indonesia. Department for International Development: TRL.

Rogers, E.M. 1983. *Diffusion of Innovations*. Fifth Edition. New York (US): The Free Press.

Saban. 2016. Bangun trotoar, Bekasi dapat penghargaan. [download 2018 Des 2]. Available at: http://pos kotanews.com/2016/11/16/bangun-trotoar-kota-bekasi-dapat-penghargaan/.

Sanders, K., & Canel, M. J. 2012. Government Communication: An Emerging Field in Political Communication Research. The sage handbook of political communication. 85–96. Available at: http://mariajo secanel.com/pdf/emergingfield.pdf.

Tamin, O.Z. 1992. Pemecahan kemacetan lalu lintas kota besar. *Journal of Regional and City Planning* 3 (4): 10-17.

[WHO] World Health Organization. 2013. *Pedestrian Safety: A Road Safety Manual for Decision-makers and Practitioners*. Geneva (CH): WHO Press.

Community Development

Rural Socio-Economic Transformation – Kinseng et al. (Eds)
© 2019 Taylor & Francis Group, London, ISBN 978-0-367-23603-8

From psychological burden to social economic changes: Analysis impact of evictions for women in Jakarta

I. Dalimoenthe, A.T. Alkhudri, R.N. Sativa, B.P. Andhyni & I. Dewi
Department of Sociology, State University of Jakarta, Jakarta, West Java, Indonesia

ABSTRACT: A 'house' does not merely mean a building to live; instead it is defined by complex and long-term relationships. House eviction could affect the psychological state as well as social-economy status— as their place turned into debris. This research is intended to analyze the effects of eviction, especially for women and children. This research is analyzed by using post-positivistic-mixed method paradigm. The research is located in DKI Jakarta and conducted for three months (March-May 2018). The samples of this research are 430 respondents, consists of women evictees in DKI Jakarta, especially DAS Ciliwung and Penjaringan residents who are relocated to Jatinegara and Marunda flats. The findings show eviction causes changes in many aspects which are directly faced by women such as; first, compulsion when relocated process, low compensation, compulsion to accept compensation, trauma, and depression. Second, dissimilar development and social exclusion. Third, social-economy changes and decreasing income.

1 INTRODUCTION

Poverty is still a problem that threatens the Capital of Jakarta. BPS DKI Jakarta stated the number of poor people in Jakarta has increased by 0.14 points since 2015. In September 2015, the number of poor people reached 368,670 people or 3.61 percent of the total population in DKI Jakarta and in early March 2016, the number of poor people increased to 384,300 people or 3.75 percent. Based on that data it means that there is an increase of 15,630 people or an increase of 0.14 points. However, when compared to March 2015 with the number of poor people amounted to 398,920 people or about 3.93 percent, then the number of poor people in March 2016 has decreased by 14,620 thousand or decreased by 0.18 points (Harian Terbit 2016).

Despite the declining in the number of poor people, poverty is still a big concern for DKI Jakarta. This happens because of the unbalanced physical development between urban and rural areas. As a result, urbanization is increasing in a massive number/over urbanization (Schumacher 1999). The effects of over-urbanization include (1) massive population growth in DKI Jakarta; (2) the high demand for residential land; (3) shifting zonation of spatial settlement; and (4) the growth of informal economy pockets (Evers 1982, Paul 1986, Santosa 1991, Kivell 1993).

These conditions create densely populated enclaves and slum areas in the river basin areas (DAS), state-owned land, and others (Sasanto & Khair 2010). The slum area is a logical consequence of the growing city. In the midst of the increasingly massive growth of the city, the social structure of the slum area community has further strengthened its position from generation to generation since 1960s (Ali & Fodmer 1969). The people adapt and survive to live in the Capital of Jakarta in all circumstances.

However, the existence of slum areas for the government is a threat to urban development. The city becomes run-down, instead of pretty. Recognizing this, the government encouraged a variety of efforts to reduce the slum areas, one of them is through eviction and relocation of the population. In practice, eviction and relocation give dire socioeconomic and political

Table 1. Various research regarding eviction in urban area.

Context	Researchers	Research Focus
Theoretical debate	Desmond and Gershenson 2016, Slater 2006, Soederberg 2018, Kahlmeter et al. 2018, Levenson 2017	Urban eviction studies are used as discourse in various theories such as urban sociology, economics, the environment, and poverty.
Development of scientific studies	Gerull 2014, Zakaree 2012, Desmond and Gershenson 2016, Bhan 2009, Dickinson 2015	Evictions are populated as studies in various scientific fields such as sociology, geography, economics, law, urban governance, and politics.
Research issue/ setting	Olds 1998, Otiso 2002, Purser 2014, Stenberg et al. 2011, Steel et al. 2014	Urban eviction studies are used in different regions or locations in the world.

consequences. These conditions triggered various studies on urban evictions, whether from the context of theoretical debates, the development of the scientific discipline, as well as the setting/issue of the study (see Table 1).

The reality of eviction changed the lives and socioeconomic orientation of the victims' families, flourished developmental injustices and social exclusion, also showed the lack of the empowerment of citizens, especially for women, mothers, and children. These consequences triggered the authors to examine the impact of evictions for women in Jakarta from psychological burden to social economic, by using structuration theory (Giddens 2010) and resilience-livelihood strategy (Chambers & Conway 1991, Crow 1989, Ellis 2000, De Haan 2000).

2 METHODOLOGY

This study used critical paradigm, mixed method - by R & D method (Borg et al. 2003, Denzin & Lincoln 2009, Lubis 2004). The research was conducted in DKI Jakarta for 3 months, from March to May 2018. The main samples of this study were family evicted from Ciliwung River and Penjaringan to Rusunawa Jatinegara and Marunda, as many as 430 respondents. The complementary subjects are the stakeholders involved in eviction and relocation, such as government officials, Social Services Agency, and NGOs.

Data collection was done by library study, survey, in-depth interview and FGD. The literature study is used in order to collect similar studies and the conceptualization of urban evictions. Related to the survey, the determination/selection of respondents was conducted by purposive sampling, while determining the number of respondents in each Rusunawa proportionally determined by considering the proportion of the adult population. In-depth interviews and FGDs were conducted with government officials, Social Services Agency, and NGOs as stakeholders interested in eviction cases. Meanwhile, data analysis relies on quantitative descriptive data analysis and qualitative analysis of interactive models (data reduction, data presentation, and conclusion) (Miles & Huberman 1994).

3 RESULTS AND DISCUSSIONS

3.1 *Description of respondents of evication victims*

Respondents in this study amounted to 430 people, which were families evicted to Ratawa Jatinegara and Marunda, DKI Jakarta. There are 51% of women and 49% of men in our samples. The religion of our respondents is Moslems (93.26%), followed by Catholic (3.02%), Protestant (3.02%), Buddhist (0.23%), Hinduism (0.23%), and others (0.23%). The average age of eviction victims was 43% of age between 31-45 years and 35% of age 45-60 years. The rest of the age of children, adolescents, to the elderly (22%).

According to the education level, 46.7% of the respondents are high school graduates, followed by middle-high school graduates(32.8%), primary school graduates (17.4%), did not go to school 0.9% and 2.1% of diploma/bachelor graduates. The high number of respondents with a low level of education, which is senior high school graduates, makes most respondents work on the informal sector, such as traders/street vendors, laborers and self-employed. (48.37%), other jobs (32.10%), and not working (19.53%). Relying on the informal sector, evictions experienced by respondents gave implications for economic vulnerability and family livelihood systems. A few shocks can make their family fall into poverty.

An eviction is a real form of shock experienced by the victim's family. The residence and the environment are not just physical things but are formed from complex relationships and long periods of time which has significant implications for family welfare. Physical space provides access to the distribution of resources needed by families such as income, knowledge, and social interaction. In fact, these conditions were experienced by most families of evicted victims who were relocated to Rusunawa Jatinegara and Marunda. They were partly as big as they are originally domiciled in the Ciliwung-Penjaringan Watershed. The dynamics of urban development, spatial arrangement, and the construction of the East Canal Flood caused them to be relocated from their homes.

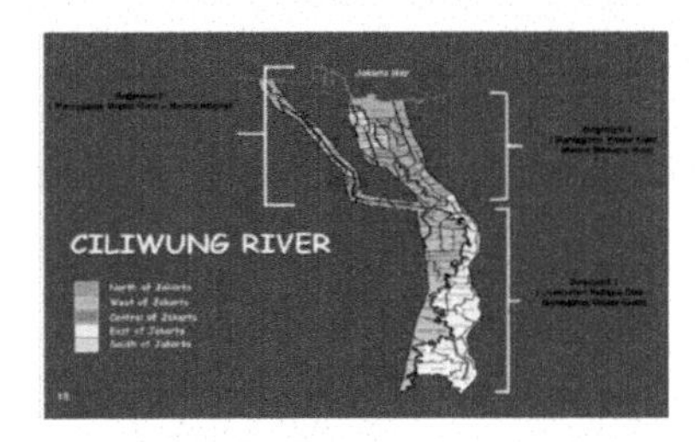

Figure 2. Map of Ciliwung River crossing DKI Jakarta Province (Source: Regional Environmental Management Agency of DKI Jakarta, Indonesia 2009).

3.2 *Social structure of families of eviction victims: pre and post eviction*

The findings reveal that after the eviction, most of the respondents are a native of DKI Jakarta, amounted to 66% and 34% and migrants. The majority of evictees have been living in evictions between 20-30 years (88%) and 10 years and over (12%). In previous locations, 83% of respondents live in their own homes, 10% live in the parents' house and another 7% rented. After the relocation to Rusunawa Jatinegara and Marunda, all respondents are tenants within a certain period. These status changes makes the people landless, vulnerable to other evictions, and suffered from the increasing cost of daily living expenses (such as leasehold, water, electricity, and food).

The increasing cost of living and the loss of livelihoods make the families evicted more vulnerable, less self-sufficient, and more dependent on government and NGO assistance. These indicate that the evictions change economic conditions (94.4%) and the orientation of the victim's life. There are 3 markers that encourage changes in life orientation, namely: loss of memory of residence, hometown, and the loss of kinship system that has been built for a long time. In the new location (Rusunawa), community solidarity is still relatively weak (42%) compared to the previous location (58%). Thus, eviction has transformed an independent and solid society into a dependent society. In short, evictions gave significant impacts on life (Table 2) and reproduce poverty to a deeper degree.

Figure 3. Description of the previous location, when evictions, and up to Rusunawa.

Table 2. Impact of Evictions.

No.	Impact of eviction	Impacted (%)	Not impacted(%)
1.	Decreased earnings	84	16
2.	Traumatic feelings	70	30
3.	Depression	42	58
4.	Exclusion	66	34

3.3 *Impacts of eviction for women and children*

Evictions often ignore deliberations between the government and citizens with inadequate compensation, thus implicating the citizens' compulsion to relocate. In the context of eviction in Ciliwung and Penjaringan basin residents, 83.2% of respondents stated that the compensation they received was not sufficient. This correlates with their high level of compulsion to move to the Rusanawa Jatinegara and Marunda sites (91.8%). Consequently eviction has always been associated with moments of injustice for the small community (74.6%) and the absence of NGOs and government. According to Giddens (2010), this indicates that the ruling structure (state) is suppressing the controlled structure (eviction victims).

> *"The compensation I received was not sufficient, it was unfair. Evictions make us suffer more. We lose jobs and fixed income."*
>
> *(Interview, May 2018)*

> *During the eviction, there are many things that we could not save. All of our belonging were missing: TV, cupboard, rice cooker, money, and more. We had no chance and strength to fight.*
>
> *(Interview, May 2018)*

The greater burden of evictions is often experienced by women, children, and the elderly. All three are susceptible to depression, trauma, and other psychological burdens. This is closely related to the heavy role in carrying out and managing the survival of the household before and after eviction. Post-eviction, family income declined. This condition drove women to think and act creatively in order to adapt or to be resilience for survival. The family evicted used resilience strategies for family livelihoods, which is cost minimization and profit maximization (Crow 1989, Ellis 2000, Chambers & Conway 1991, de Haan 2000, Ellis 2000), where the victims chose five conditions: (1) for poor families with unemployed husbands, they seek new jobs that suit their abilities; (2) for poor families who still pursue their own business will apply for side jobs to earn additional income; (3) empower family members to seek additional income; (4) tightening consumption by reducing the quality of the food menu, where this choice is usually done by women; and (5) perform the mechanism of "dig hole closing hole", a way of pay off debt with another debt.

In terms of accessibility, the eviction made access to daily activities harder for the victims, where they were forced to live far away from the market/center of merchants, factories, offices and even to the school for the children. For example, a mother usually accompanied their children to go to school, but the eviction made the children stopped going to school since no one could accompany them to school.

> *My children no longer go to school, because I can no longer accompany them.... The distance is too far, the cost is expensive. Not to mention, I have to help the family by looking for side jobs.*
>
> *(Interview, May 2018)*

From the description above, it is understood that eviction increased the burden and role of women in the household. These impacts drove women to be more resilient for the sake of family livelihood. To help women as victims of eviction, empowerment models and programs could be designed to support the families evicted. This is important so that they can be more resilient, independent, out of poverty and out of social exclusion. To realize the right empowerment model, embedding psychoeducation program and strengthening the instituion of Rusunawa communities are needed.

4 CONCLUSIONS

Eviction alters the orientation of life, gives psychological pressure to the victim's family (trauma and depression), and lower their social-economy status. These conditions are majorly suffered by women, children, and the elderly. Therefore, displacing and relocating residents to Rusunawa might be not the best solution. Relocation should have a humanitarian approach by considering the complexity of people's life. Consideration of human character, socio-culture, region/physical, and source of livelihood have to be seen as priorities for relocation cases. These four aspects are important to consider as parameters for relocation policy decision making. In addition, psychoeducation is urgently needed, especially to empower women and children sustainably. Other than psychoeducation program, training program such as management of livelihood in relocation sites is also needed especially for women. This program is important because women have a significant role in supporting the family. This training program could also be strengthened by empowering the victim based on the local wisdom in Jatinegara and Marunda Rusunawa.

ACKNOWLEDGEMENT

We express our highest gratitude to the field assistants and respondents and key informants in the two research locations, Marunda and Jatinegara Barat Flats. On this occasion, we would also like to thank the Ministry of Research and Higher Education, Republic of Indonesia, which has provided basic research grants to universities in the 2018 budget year through State

University of Jakarta. The paper will not be present without the assistance of the research grant. This paper was written with financial support from the basic university research grant scheme, the ministry of research, technology and higher education facilitated by Institute for Research and Community Service *(Lembaga Penelitian dan Pengabdian kepada Masyarakat)*, Jakarta State University, 2018.

DISCLOSURE STATEMENT

This paper has no potential conflict of interest and politics. This paper was compiled by the author as an academic narrative of social problems experienced by urban communities, especially those affecting women.

REFERENCES

Ali, R. M. & Bodmer, F. 1969. Djakarta Djaja Sepandjang Masa. Jakarta: Pemerintah Daerah Khusus Ibukota Djakarta.

Bhan, G. 2009. 'This Is No Longer the City I Once Knew': Evictions, the Urban Poor and the Right to the City in Millennial Delhi. *Environment and Urbanization* 21(1):127-142.

Borg, W. R., M. D. Gall, & J. P. Gall. 2003. *Educational Research an Introduction.* 7th ed. Boston: Pearson Education, Inc.

Chambers, R. & G. R. Conway. 1991. *Sustainable Rural Livelihoods and Landscapes: Practical Concepts for the 21st Century.* Sussex.

Crow, G. 1989. The Use of the Concept of 'strategy' in Recent Sociological Literature. *Sociology* 23 (1):1–24.

Denzin, N. K. & Y. S. Lincoln. 2009. *Handbook of Qualitative Research.* 2nd ed. Thousand Oaks, CA: SAGE Publications.

Desmond, M. & Gershenson, C. 2016. Who Gets Evicted? Assessing Individual, Neighborhood, and Network Factors. *Social Science Research* 62:362–77.

Dickinson, G.S. 2015. Towards a New Eviction Jurisprudence. *Georgetown Journal on Poverty Law Policy* 23(1):1–59.

Ellis, F. 2000. *Rural Livelihoods and Diversity in Developing Countries.* Oxford: Oxford University Press.

Evers, Ha. 1982. *Urban Sociology-Urbanization and Land Dispute in Indonesia and Malaysia.* Jakarta: LP3ES.

Gerull, S. 2014. Evictions Due to Rent Arrears : A Comparative Analysis of Evictions in Fourteen Countries. *European Journal of Homelessness* 8(2):137–55.

Giddens, A. 2010. *Sociology.* 6th ed. Maiden: Polity Press.

De Haan, L. J. 2000. Globalization, Localization, and Sustainable Livelihood. *Sociologia Ruralis* 40 (3):339–365.

Harian Terbit. 2016. Era Ahok Jumlah Orang Miskin Di DKI Meningkat. *Harian Terbit.* Retrieved from http://megapolitan.harianterbit.com/megapol/2016/07/19/65956/18/18/Era-Ahok-Jumlah-Orang-Miskin-Di-DKI-Meningkat at 28 March 2018.

Kahlmeter, A., Bäckman, O., & Brännström, Lars. 2018. Housing Evictions and Economic Hardship. A Prospective Study. *European Sociological Review* 34(1):106–19.

Kivell, P. 1993. *Land and the City: Pattern and Process of Urban Change.* New York: Routledge.

Levenson, Za. 2017. A Wake for Urban Theory Ghetto: The Invention of a Place, the History of an Idea Evicted: Poverty and Profit in an American City. *The British Journal of Sociology* 68(4):785-790.

Lubis, A. 2004. *New Paradigm and Issues on the Methodology of Social Sciences-Humanities and Culture in the Postmodern Era.* Jakarta: PPS UI.

Miles, M.B. & Huberman, A. M. 1994. *Qualitative Data Analysis: An Expanded Sourcebook.* Thousand Oaks: Sage.

Olds, K. 1998. Urban Mega-Events, Evictions and Housing Rights: The Canadian Case. *Current Issues in Tourism* 1(1):2-46.

Otiso, K.M. 2002. Forced Evictions in Kenyan Cities. *Singapore Journal of Tropical Geography* 23 (3):252-267.

Paul, B.K. 1986. Urban Concentration in Asian Countries : A Temporal Study. *Journal Area* 18 (4):299–306.

Purser, G. 2014. The Circle of Dispossession: Evicting the Urban Poor in Baltimore. *Critical Sociology* 36(1):45-64.

Regional Environmental Management Agency of DKI Jakarta, Indonesia 2009. The Report of RT/RW Arrangement for DKI Jakarta 2030. Jakarta: BAPPEDA.

Santosa, I. 1991. *Overview of Homelessness in Industrial Cities and Non-Industrial Cities: Case Studies in Yogyakarta and Semarang*. Jakarta: Toyota Foundation and Yayasan Ilmu-Ilmu Sosial Indonesia.

Sasanto, R. & Khair, A.S. 2010. Analisis Kebijakan Pemerintah Dalam Penanganan Permukiman Ilegal Di Bantaran Sungai Studi Kasus: Bantaran Kali Pesanggrahan Kampung Baru, Kedoya Utara, Kebon Jeruk. *Jurnal Planesa* 1(2):146–52.

Schumacher, E. F. 1999. *Small Is Beautiful: Economics as If People Mattered : 25 Years Later . . . with Commentaries*. Point Roberts, Wash: Hartley & Marks Publisher.

Slater, T. 2006. The Eviction of Critical Perspectives from Gentrification Research. *International Journal of Urban and Regional Research* 30(4):737–57.

Soederberg, S. 2018. Evictions : A Global and Capitalist Phenomenon. *Development* 49(2):1–24.

Steel, W.F., Ujoranyi, T.D., & Owusu, G. 2014. Why Evictions Do Not Deter Street Traders: Case Study in Accra, Ghana. *Ghana Social Science Journal* 11(2):52-76.

Stenberg, S.A., Van Doorn, L., & Gerull, S. 2011. Locked out in Europe : A Comparative Analysis of Evictions Due to Rent Arrears in Germany, the Netherlands and Sweden. *European Journal of Homelessness* 5(2): 39-61.

Zakaree, Saheed. 2012. Externalities of Urban Redevelopment: Eviction, Relocation, and Compensation in Nigeria. *International Journal of Business and Social Science* 3(5):272–78.

Rural Socio-Economic Transformation – Kinseng et al. (Eds)
 ISBN 978-0-367-23603-8

Social resilience of the farmers community to cope with climate change

I. Wulansari
Postgraduate program in Department of Sociology, Padjadjaran University, Bandung, West Java, Indonesia

O.S. Abdoellah & B. Gunawan
Institute of Ecology and Department of Anthropology, Padjadjaran University, Bandung, West Java, Indonesia.

Parikesit
Institute of Ecology and Department of Biology, Padjadjaran University, Bandung, West Java, Indonesia

ABSTRACT: Climate change poses a serious vulnerability to the agricultural sector in Indonesia, where farmers community needs resilience to cope with the problem. We have identified farmers' resilience through in-depth interviews and participant observation. Empirically, in one village namely Nunuk Village in Indonesia there are three things, namely climate change vulnerability, potential for transformation and farmers' adaptive response. First, vulnerability is characterized by the occurrence of more than 20 years of drought and brown plant hopper attacks in the past 10 years. Second, transformability is indicated by the potential for transformation marked by agents of farmers who practice the principle of sustainability agriculture. Third, adaptability is characterized as an adaptive response to farmers community that is demonstrated through a mechanism of timing of planting rice collectively. There is a room for improvement, that the resilience of farmers community in facing climate change needs crisis management mechanism.

1 INTRODUCTIONS

1.1 *The impact of climate change*

Climate change poses a serious vulnerability to the agricultural sector. Indonesia is a country prone to climate change because economic activities have a high dependence on the agricultural sector (Syaukat 2011). In Indonesia, climate change is in the form of El Nino Southern Oscillation (ENSO) which has an impact on annual variations in the rainy season caused by Austral-Asian monsoon (Naylor et al. 2007). In addition, climate anomalies have the effect of decreasing crop production or even resulting in crop failure due to lack of water resources. In order to defend themselves against vulnerability, farmers need resilience or resilience to face pressure. One of the resilience of farmers is facing water scarcity.

The resilience of farmers facing the water crisis needs to be supported by government policies. Resilience is reinforced by government policies as in China through the Water Saving Irrigation (WSI) program which contributes positively to the reduction of greenhouse gases and saves water by 30 Bm3 per year (Zou et al. 2012). In addition, farmers' resilience is determined by the existence of irrigation infrastructure as an adaptation effort that can secure and store water (Burney et al. 2014). Adaptation efforts do not stand alone because the act of adaptation of farmers first begins with the belief of farmers in the occurrence of climate change. This is based on research by Belay et al. (2017) in Central Rift, Ethiopia that if farmers' awareness of current climate change will be followed by adaptation strategies including crop diversification, adjustment of planting time, soil and water conservation and management and diversification of income through livestock activities.

Whereas in Malaysia, farmers need cooperation with government and non-governmental institutions to make appropriate adaptations to reduce vulnerability due to climate change (Masud et al. 2017). Connectivity with government and non-governmental institutions can be one of the implications of adaptive capacity (Adger 2001). However, the high and low adaptive capacity can be measured from economic resources, awareness of climate change and the technological capacity of farmers (Abdul-Razak & Kruse 2017). Another thing that causes low adaptive capacity is that there is no collective action in livelihood activities (Panjaitan et al. 2016).

1.2 *The concept of social resilience*

This paper focuses on the concept of social resilience. The strongest reason for using the concept of social resilience is the affirmation of Carmin et al. (2015) regarding the phenomenon of climate change raises vulnerability to individuals and groups, especially those with limited capacity to overcome various pressures. Climate change raises uncertainties in climate conditions, increases weather and seasonal variability, increases the frequency of climate events (Tompkins & Adger 2004). Folke et al. (2003) provide reinforcement that building resilience is a learning effort to live in various changes and uncertainties.

The concept of social resilience is raised by Adger (2000) who reviews social groups or communities that depend on ecological resources and the environment as their livelihood. Then social resilience is defined as the ability of groups or communities to overcome external pressures and disturbances as a result of social, political and environmental changes. Whereas Kwok et al. (2016) define social resilience as a community capacity that contributes to community readiness to respond to disasters and post-disaster recovery.

In a study of fishing communities, Pauwelussen (2010) emphasized that the definition of social resilience is strength or cohesion at the community level with the approach of actors and groups of actors involved in the social environment. Social resilience is closely related to the daily practices of social actors, social networks and the power of actor agencies (Sakdapolrak et al. 2016). Social resilience lies in coping strategies that rely on the ability of the actor to overcome difficulties (Keck & Sakdapolrak 2013). The difficulties faced by actors require knowledge to enable actors to come out of pressure. Knowledge is property resilience that can help manage and use resources to maximize sustainability and minimize negative impacts (Maclean et al. 2014). Haque et al. (2017) and Alam et al. (2017) in Bangladesh confirms that knowledge can determine the basis of climate change adaptation criteria and support farmers' adaptive capacity.

However, how enrichment of knowledge can persistently strengthening of farmers' capacity and voluntary acceptance as a practice for farmers in shaping resilience. Berkes and Turner (2006) state that practice is based on self-interest that carries a variety of norms. Practices based on the knowledge and awareness of farmers do not always determine the decision-making process for farmers (Vanclay & Lawrence 1994). In making decisions, farmers act as producers who consider the use of synthetic pesticides as a solution to overcome pests (Smit & Skinner 2002, Wati & Chazali 2015). The use of chemical pesticides is a threat to agriculture because it has the potential to destroy sustainable agriculture (Antwi-Agyei et al. 2017). Sustainable agriculture is the solution to climate change (Zougmore et al. 2016). Climate change is still a 'discourse' for farmers so that there is a gap between farmers' practices and knowledges in general, referring to productivity and practices that farmers should do so that livelihood sustainability and agricultural ecosystems occur.

2 METHODS

2.1 *Research method*

This research uses a qualitative method to build a relationship between the concept of social resilience which relies on the capacity of farmer agents and property in agriculture which is commonly called an agroecosystem. Marten (1988) explained that in the concept of agroecosystem there are two properties, namely productivity and sustainability. Productivity can be negatively related to sustainability. High productivity can be a low sustainability if production using solid inputs (one of which is pesticide use) can lead to ecosystem changes that ultimately weaken production.

In addition, the present study shows the adaptive response of farmers' community in dealing with climate change. This is seen from the concept of Resilience Thinking (Folke et al. 2010). Two aspects of Resilience Thinking include adaptability and transformability which are the basis of analysis of how the adaptive response of the farmer's community in facing climate change.

Primary data collection techniques through in-depth interviews and observations in the period from May to December 2018. Primary data obtained directly from informants determined purposively. The idea of qualitative research is to ensure participants are deliberately chosen (purposefully) that can help researchers understand the problem and research questions (Creswell 2009).

Determination of informants is based on the consideration that the selected person has the information needed by the researcher. In this study, data were taken from the following sources: group of farmers, Farmer Field School alumni, Farmer Rainfall Observers Club, village government apparatus, agricultural extension agencies and the Regency Agricultural Agency. These informants can represent an explanation of the vulnerability and adaptive response of farmers to climate change.

2.2 *Study area*

This research was conducted in Nunuk Village, which is one of the 11 villages in the sub-district of Lelea, Indramayu Regency, West Java Province, Indonesia. The people of Nunuk generally have farming livelihoods both as farm labourers, owners and cultivators as well as owners of agricultural land. Generally, the ownership of paddy fields in the Nunuk Village ranges from 200 bricks (28 square meters) to 1 hectare. Data from the Nunuk Village Hall that the work of farmers and farm labourers occupies the majority occupation where the population working as farmers reach 1,275 people and farm workers reach 1,932 people.

Based on the results of the study of Estiningtyas et al. (2011) that Lelea sub-district (where the Nunuk Village is located) is one of four sub-districts in Indramayu Regency that are vulnerable to climate anomalies because they are vulnerable to drought. Data from the Irrigation Service Office of Indramayu Regency shows that Lelea sub-district faces low criteria of rainfall in the period 1987–2017. Within 30 years, Lelea Subdistrict faces drought due to low rainfall (0–100 mm/month) for 22 years with a cycle that often occurs which is 3 months without rain.

Another reason for choosing Nunuk Village is because it has indications of community resources. Resources are the main thing in social resilience. Social resilience is related to the ability of the community to access important resources (Langridge et al. 2006). First, there are key individuals who have leadership as vital community resources (Folke et al. 2005, Longstaff et al. 2010). Second, there is innovative learning in social groups informally (Longstaff et al. 2010). Third, having community networks that become supporters when communities face crises and events (Maclean et al. 2014). Fourth, communities have the potential for transformation shown by using information and knowledge for adaptation strategies to avoid repeating mistakes (Folke 2006, Smit & Wandel 2006).

3 RESULTS

3.1 *Pests in paddy field*

Climate change and the lack of natural predators to prey on insects and mice are serious pest problems in the Nunuk Village. The pest that most worries farmers are Brown Planthopper or

Nilaparvata lugens Stal. Brown planthopper is an insect that can multiply rapidly which perch on rice stems. Even the brown planthopper population can explode to be able to produce 350 eggs in just 2 weeks during the rainy season (Fox 2016). In the 10 years, the population of brown planthopper has increased in the Nunuk Village. Most of the farmers in Nunuk stated that brown planthopper developed rapidly when there was rain during the day which caused high humidity. The brown planthopper attack in Nunuk Village occurred in the 2016/2017 planting season. Farmers in the Nunuk Village experienced a drastic decline in yields. By illustration, normal yields for 100 bricks (per 14 square meters) reach 9 quintals, but yields in the second 2016/2017 growing season for 100 bricks only reach 2 quintals. According to Indramayu Regency Central Bureau of Statistics data, the area of Lelea sub-district which was attacked by brown planthopper was 104 ha in 2016.

Farmers in Nunuk Village generally grow rice twice a year. While some of Nunuk's neighbouring villages grow rice three times a year. So, planting rice without a pause, planting three times a year causes the development of the brown planthopper to be fertile. In addition, brown planthopper migration occurred in the third planting season from the neighbouring village to Nunuk Village. The brown planthopper life cycle is not interrupted even during 2017, Nunuk farmers enter a period of sadness because the yields have dropped dramatically. As a result of the brown planthopper migration, the farmers' fields were burnt and rice grassy stunt virus and rice ragged stunt virus were burned.

In the first planting season of 2018 (May to August 2018) rainfall in Nunuk Village is quite low. Based on the observation of the Rainfall Measuring Station from Tugu Station which is the closest from Nunuk Village that in May there are 3 rainy days with a total of 36 millimetres, in July there are 2 rainy days with an amount of 21 millimetres. Although throughout 2018 there tends to be low rainfall, yields are better than in 2017. In comparison, according to data from the Indramayu District Agriculture Office that during the two planting seasons in 2017, the average yield in Lelea sub-district is around 35,000 tons, while in two 2018 planting season increased in the first season to reach 39,000 tons, while in the second season reached 36,000 tons. Generally, according to farmers in the Nunuk Village in 2018, the harvest is relatively close to normal at an average of 7 tons per hectare.

3.2 *Use of synthetic pesticides*

The principle that is still adopted by farmers in the Nunuk Village is the handling of brown planthopper pests through the application of synthetic insecticides. Even eloquent farmers mention the name of an insecticide product that is famous for being expensive and effective in expelling brown planthopper with the local term namely *dekles*. *Dekles* means killing pests directly, including natural predators of pests. In addition to the use of expensive synthetic insecticides, farmers also mix chemicals to kill the brown planthopper. Opposition material consists of a mixture of kerosene, chlorine and washing soap. Not only types of mixing, in the planting season in 2017, but some farmers also apply synthetic pesticides up to 8 times in 1 planting season. In fact, there is a farmer applying synthetic pesticides up to 10 times in one planting season.

Generally, farmers apply insecticides without making observations so that insecticide use is not on target. This is contrary to the principle of Integrated Pest Management that has been studied by farmers in Farmer Field School. According to the principle of Integrated Pest Management, the application of pesticides can be carried out after by observing the economic threshold marked by the discovery of 40 brown planthoppers in 1 clump of rice plants. In Farmer Field School, trials of land that did not use pesticides and used pesticides as a comparison were practised. Yields on land that does not use pesticides are far greater than land that uses pesticides. Although it has been proven that not using synthetic pesticides as long as pests are still at a reasonable threshold it will not affect the yield, but the practice of farmers is difficult to reduce the use of pesticides. Farmers of Farmer Field School alumni no longer practice the knowledge gained. One factor is community pressure that makes farmers not maintain behaviour-based Integrated Pest Management. Farmers of Farmer Field School alumni often become a mockery because they are considered unable to work on rice fields because of the poor appearance of their fields.

3.3 *Potential for transformation towards sustainable agriculture*

Agents of farmers who are members of Farmer Rainfall Observers Club are able to show 'differences' in farmers community. The agents believe that the benefits of observations based on Integrated Pest Management Principle on their land can help to minimize the use of synthetic pesticides. The Farmer Rainfall Observers Club is a form of institutionalization of learning through the Science Field Shops. Among all farmer field schools in Nunuk Village, Science Field Shops is the longest learning activity in Nunuk Village for up to 8 years (2010–2018). Science Field Shops is an agrometeorology learning that becomes the learning of farmers facing climate change by implementing the formation of the club of Indramayu Farmer Rainfall Observers Club. The participation of farmers in the Farmer Rainfall Observers Club is characterized by the obligation to measure rainfall every morning and the obligation to make observations and records on land and rice plants. Another obligation is monthly evaluation among group members.

While the Nunuk Village has the potential for transformation, the term climate change is still not fully believed by farmers. The terms of climate change such as La Nina and El Nino are considered as the terminology of languages that are not grounded for farmers because the term is more familiar to academics. There is a gap between agents of Farmer Rainfall Observers Club and the general farming community. The terms El Nino and La Nina in the terminology of climate change are still limited to discourse for the farming community in general, so there is no awareness to change daily practices to support adaptation to climate change. Furthermore, agents of Farmers Rainfall Observers Club believe that climate change is a real threat to survival as a farmer and thus requires a change in behaviour. Behaviour change is shown by improving farming practices that are more adaptive to climate change.

3.4 *Adaptation through determination of collective planting time*

Although climate change is not yet fully understood by farming communities in Nunuk village, but institutionally Nunuk village has adaptive capacity. This is indicated by the strategy of determining the planting time through village meetings before the planting season which is carried out twice a year. Deliberations are held if there has been a decision from the authority handling irrigation regarding the time of irrigation rotation which indicates the adequacy of water to start the time of planting rice. The meeting became a tradition institutionalized by the Nunuk village apparatus since 1998. The tradition began after the end of the Integrated Pest Management training program known as Farmer Field School. The Farmer Field School is a national program from the central government with technical assistance from FAO (Food Agricultural Organization) to rice farmers (Winarto 1999). The training program took place because in the period 1994 to 1996, a group of white rice stemborer (Scirpophaga incertulatas) occurred in Indramayu Regency and Nunuk Village was exposed to these pests.

In the Farmer Field School, farmers learn the formula of avoiding white rice stemborer known in local terms as a *genggong* formula. The village deliberation determined the timing of planting through the *genggong* formula. The *genggong* formula is an instrument to anticipate the planting time so that it does not collide with the peak of the flight of white rice stemborer. The village deliberation was facilitated by the Nunuk village apparatus which was attended by representatives of regional governments in the field of irrigation, agricultural extension agencies, representatives of farmer groups and farmers' communities. As for the formula, the calculation of the *genggong* formula is presented by farmer who are alumni of the Farmer Field School and chair of the Farmer Rainfall Observers Club. In addition, in the deliberations for planting time, the presence of the Farmer Rainfall Observers Club reads out seasonal scenarios as a precaution to choose the appropriate planting time. Farmers who are members of the Farmer Rainfall Observers Club get a seasonal scenario from scientists (anthropologist and agrometeorologist). These scientists built scientific collaboration through agrometeorological learning to farmers through the Science Field Shops arena.

4 DISCUSSIONS

Yields are a measure of productivity. This is in line with the concept of agroecosystem from Marten (1988) that productivity contributes to the stability of farmers' income. However, agricultural productivity is under threat due to the interrelationship of temperature, rainfall and extreme events causing the spread of pests and diseases in plants (Lobell & Burke 2008). In handling pests, the Indonesian government officially chooses pesticides or eradicates chemical pests (Casson 2016). The use of synthetic pesticides is part of a modern technology to support the stability of agricultural production (Marten 1988). However, farmers need to improve agricultural practices, one of which is the use of pesticides to be able to adapt to climate change (Belay et al. 2017). Sustainable agricultural practices and strengthened by the ability of communities will encourage social resilience to face climate change (Altieri et al. 2015).

Facing the effects of climate change requires farmers' responses and adaptation strategies for agricultural production (Lobell & Burke 2008). In the face of climate change, farmers need the ability to deal with climate uncertainty and strategies to maintain their livelihoods. In this case, adaptation needs to be measured in terms of other alternative views, namely ability to achieve resilience (Nelson et al. 2007). The concept of resilience proceeds through the idea of adaptive capacity and the ability of the social system to learn and adapt to disruptions (Folke et al. 2005). Furthermore, Folke et al. (2010) introduced the concept of resilience thinking with three attributes, namely resilience, adaptability and transformability. Adaptability is a resilient part that represents the action capacity of adjusting to change external drives and internal processes. Transformability is a transformed capacity in stability landscape that can create a new system that is fundamentally ecological, economical or through social structures.

Adaptability is demonstrated by the adaptation strategies carried out by the community in the face of external pressure. One of the adaptation strategies undertaken by farmers is to change the planting calendar (Belay et al. 2017). The adaptation strategy is obtained by the farmer community through learning and applying new knowledge. Learning to adapt to changes and uncertainties combined with various types of knowledge for learning can form adaptive capacities (Folke et al. 2003). Both forms of training such as the Farmer Field School and the Science Field Shops provide knowledge to farmers. Nunuk Village is an innovative learning arena. Communities can learn and innovate through innovative learning both individually and in groups through trial and error experiments (Longstaff et al. 2010).

5 CONCLUSIONS

Resilience of the farmers' community to face vulnerability due to climate change can be overcome through adaptability. Adaptability is demonstrated by the knowledge-based adaptive response of the farmers' community. However, community resources cannot work effectively if they are not accompanied by a joint crisis resolution mechanism. If the farming community has a strategic mechanism to start the planting period into an adaptation strategy, it needs to be equipped with a mechanism to deal with the crisis. This is due to the vulnerability that has the potential to reduce crop yields due to the spread of pests in rice plants. Furthermore, resilience is not only a form of new adaptation but needs to be a form of change in farming practices that support livelihood sustainability and ecosystem sustainability.

REFERENCES

Abdul-Razak, M. & Kruse, S. 2017. The adaptive capacity of smallholder farmers to climate change in the Northern Region of Ghana. *Climate Risk Management* 17: 104–122.

Adger, W.N. 2000. Social and ecological resilience: Are they related? *Progress in Human Geography* 24 (3): 347–364.
Adger, W.N. 2001. Social capital and climate change. Tyndall Centre for Climate Change Research. Norwich: University of East Anglia. Working paper 8: 4–20.
Altieri, M.A., Nicholls, C.I., Henao, A., Lana, M.A. 2015. Agroecology and the design of climate change-resilient farming systems. *Agronomy Sustainable Development* 35: 869–890.
Alam, G.M.M., Alam, K., Mushtaq, S. 2017. Climate Change Perceptions and Local Adaptation Strategies of Hazard-prone Rural Households in Bangladesh. *Climate Risk Management* 17: 52–63.
Antwi-Agyei, P., Dougill, A.J., Stringer, L.C., Codjoe. 2018. Adaptation opportunities and maladaptive outcomes in climate vulnerability hotspots of northern Ghana. *Climate Risk Management* 19: 83–93.
Belay, A., Recha, J.W., Woldeamanuel, T., Morton, J. F. 2017. Smallholder farmer`s adaptation climate change and determinants of their adaptation decisions in the Central Rift Valley of Ethiopia. *Agriculture & Food Security* 6: 1–13.
Berkes, F. & Turner, N.J. 2006. Knowledge, learning and the evolution for conservation practice for Social-Ecological System Resilience. *Human Ecology* 34: 479–494.
Burney, J., Cesano, D., Russel, J., La Rovere, E.L., Corral, T., Coelho, N.S., Santos, L. 2014. Climate change adaptation strategies for smallholder farmers in the Brazilian Sertao. *Climatic Change* 126: 45–59.
Carmin, J., Tierney, K., Chu E., Hunter, L.M. 2015. Adaptation to climate change. In: Dunlap, R.E., Brulle, R.J. (eds), *Climate change and society sociological perspectives*:164–198. New York: Oxford University Press.
Casson, S.A. 2016. Ensuring community and agricultural resilience to climate change: ceremonial practices as emic adaptive strategies. *Future of Food: Journal on Food, Agriculture and Society* 4: 19–30.
Creswell, J.W. 2009. *Research Design Qualitative, Quantitative and Mixed-Methods Approaches*. California: Sage Publications.
Estiningtyas, W., Boer, R., Las, I., Buono, A., Rakhman, A. 2011. Climate risk delineation and evaluation of rainfall and rice production relationship models in supporting the development of climate index insurance in rice-based farming systems. *Jurnal Ilmu Pertanian Indonesia (Journal of Indonesian Agricultural Sciences)* 16: 198–208.
Folke, C., Colding, J., Berkes, F. 2003. Synthesis: Building resilience and adaptive capacity in Social-Ecological Systems. In: Berkes, F., Colding, J., Folke, C. (eds), *Navigating Social-Ecological System. Building Resilience for Complexity and Change*: 252–287. New York: Cambridge University Press.
Folke, C., Hahn, T., Olson, P., Norberg, J. 2005. Adaptive governance of Social-Ecological Systems. *Annual Review Environment Resources* 30: 441–473.
Folke, C. 2006. Resilience: The emergence of a perspective for Social-Ecological System analyses. *Global Environmental Change* 16: 253–267.
Folke, C., Carpenter, S.R., Walker, B., Scheffer, M., Chapin, T., Rockstrom, J. 2010. Resilience Thinking: Integrating Resilience, Adaptability and Transformability. *Ecology and Society* 15. https://www.ecologyandsociety.org/vol15/iss4/art20/
Fox, J.J. 2016. Rapidly breeding insects are threatening rice production in Java. In: Winarto, Y.T. (eds), *Krisis Pangan dan "Sesat Pikir": Mengapa Masih Berlanjut? (Food Crisis and "Misguided Thinking": Why Does It Continue?)*:41–44. Jakarta: Yayasan Pustaka Obor Indonesia.
Haque, M.M., Bremer, S., Bin Aziz, S., Van Der Sluijs, J.P. 2017. A Critical Assessment of Knowledge Quality for Climate Adaptation in Sylhet Division, Bangladesh. *Climate Risk Management* 16: 43–58.
Keck, M. & Sakdapolrak, P. 2013. What is Social Resilience? Lesson Learned and Ways Forward. *Erdkunde* 67: 5–19.
Kwok, A.H., Doyle, E.E.H., Becker, J., Johnston, D., Paton, D. 2016. What is 'social resilience'? Perspectives of disaster researchers, emergency management Practitioners and policymakers in New Zealand. *International Journal of Disaster Risk Reduction* 19: 197–211.
Langridge, R., Christian-Smith, J., Lohse, K.A. 2006. Access and Resilience: Analyzing the Construction of Social Resilience to the Threat of Water Scarcity. *Ecology and Society* 11 (2):18.
Lobell, D.B., Burke, M.B. 2008. Why are Agricultural Impacts of Climate Change so Uncertain? The Importance of Temperature Relative to Precipitation. *Environmental Research Letters* 3: 1–8.
Longstaff, P.H., Amstrong, N.J., Perrin, K., Parker, W.M., Hidek, M.A. 2010. Building Resilient Communities: A Preliminary Framework for Assessment. *Homeland Security Affairs* VI: 1–23.
Maclean, K., Cuthill, M., Ross, H. 2014. Six Attributes of Social Resilience. *Journal of Environmental Planning and Management* 57: 144–156.
Marten, G.G. 1988. Productivity, Stability, Sustainability, Equitability and Autonomy as Properties for Agroecosystem Assessment. *Agricultural Systems* 25: 291–316.

Masud, M.M., Azam, M.N., Mohiuddin, M., Banna, H., Akhtar, R., Alam, A.S.A.F., Begum, H. 2017. Adaptation barriers and strategies towards climate change: Challenges in the agricultural sector. *Journal of Cleaner Production* 156: 698–706.
Naylor, R.L., Battisti, D.S., Vimont, D.J., Falcon, W.P., Burke, M.B. 2007. Assessing risks of climate variability and climate change for Indonesian rice agriculture. *The National Academy of Sciences of the USA PNAS* 104 (19): 7752–7757.
Nelson, D.R., Adger, W.N., Brown, K. 2007. Adaptation to environmental change: Contributions of a resilience framework. *The Annual Review of Environment and Resources* 32: 395–419. https://doi.org/10.1146/annurev.energy.32.051807.090348
Panjaitan, N.K., Adriana, G., Vitrianita, R., Karlita, N., Cahyani, R.I. 2016. Adaptive capacity of coastal community to food insecurity due to climate change-a case of village in West Java. *Sodality* 4: 281–290.
Pauwelussen, A. 2016. Community as network: exploring a relational approach to social resilience in coastal Indonesia. *Maritime Studies* 15: 2–19.
Sakdapolrak, P., Naruchaikusol, S., Ober, K., Porst, S.P.T., Rockenbauch, T., Tolo, V. 2016. Migration in changing climate. Toward a Translocal Social Resilience Approach. *DIE ERDE Journal of the Geographical Society of Berlin* 147: 81–94.
Smit, B. & Skinner, M.W. 2002. Adaptation options in agriculture to climate change: A Typology. *Mitigation and Adaptation Strategies for Global Change* 7: 85–114.
Smit, B. & Wandel, J. 2006. Adaptation, adaptive capacity and vulnerability. *Global Environmental Change* 16: 282–292.
Syaukat, Y. 2011. The impact of climate change on food production and security and its adaptation programs in Indonesia. *J. ISSAAS* 17: 40–51.
Tompkins, E.L. & Adger, W.N. 2004. Does adaptive management of natural resources enhance resilience to climate change. *Ecology and Society* 9 (2): 1–14.
Vanclay, F., Lawrence G. 1994. Farmer rationality and the adoption of environmentally sound practices; A critique of the assumptions of tradititional agricultural extension. *European Journal of Agricultural Education and Extension* 1: 59–90.
Wati, H., Chazali, C. 2015. Indonesia's rice farming system in the perspective of social efficiency. *Jurnal Analisis Sosial (Journal of Social Analysis)* 19: 41–56.
Wahyono, A., Imron, M., Nadzir, I. 2014. Resilience of Fishermen Community in Facing Climate Change: Case in Grajagan Pantai Village, Banyuwangi, East Java. *Jurnal Masyarakat dan Budaya (Journal of Society and Culture)* 16: 259–274.
Winarto, Y.T. 1999. Dari paket teknologi ke prinsip ekologi: Perubahan pengetahuan Petani tentang pengendalian hama (From the technology package to ecological principles: Changes in Farmers' knowledge of pest control). In: Adimihardja, K. (eds), *Petani Merajut Tradisi Era Globalisasi* (Farmers, Linking Tradition with the Globalization Era):181–202. Bandung: Humaniora Utama Press.
Zou, X., Li, Y., Gao, Q., Wan, Y. 2012. How water saving irrigation contributes to climate change resilience- A case study of practices in China. *Mitigation Adaptation Strategy Global Change* 17:111–132.
Zougmoré, R., Partey, S., Ouédraogo, M., Omitoyin, B., Thomas, T., Ayantunde, A., Ericksen, P., Said, M., Jalloh, A. 2016. Toward climate-smart agriculture in West Africa: a review of climate change impacts, adaptation strategies and policy developments for the livestock, fishery and crop production sectors. *Agriculture & Food Security* 5:1–16.

Rural Socio-Economic Transformation – Kinseng et al. (Eds)
© 2019 Taylor & Francis Group, London, ISBN 978-0-367-23603-8

Roles of fisher folk social organization in Pati Regency

W.H. Situmeang, C.N. Nasution, A.U. Seminar & R.A. Kinseng
Department of Communication and Community Development Sciences, Bogor Agricultural University, Bogor, West Java, Indonesia

ABSTRACT: Fisher folk organizations in the past has very limited roles, however, with the development of democracy in Indonesia, the roles of fisher folk social organizations began to be more significant. This research was conducted on the fishing community in Pati, Juwana, Central Java, which consisted of various kinds of fishing sub-communities with various fishing gear, such as purse seine, cantrang, squid nets, and nets. This study aims to look at the social organization of fishers in Pati and the role of fisher folk organizations. The methodology used in this study was a qualitative approach with in-depth interviews on actors who play a role in fisher organizations and members of the organizations. Research shows that there are several and diverse fisher folk organizations. The roles of fisher folk social organizations are also very diverse, ranging from internal roles to external roles that shape relationships between organizations, including initiating fisher folk social movement.

1 INTRODUCTION

The Republic of Indonesia consists about 75% of the sea, so that it is known as a maritime country. In line with that, Indonesia has more than 17,000 island with long line beach reaching 95,181 km. Hence, it's not surprising if many Indonesian residents have a livelihood that came from or related with the sea. One of the community that depends on it is the fisher folk. The statistics of fishery shows that in 2014 the number of fisher folk in Indonesia are 2,74 million people (Ministry of Maritime and Fisheries Affairs 2015).

Most of the fisher folk in Indonesia are classified as small fisher folk. As served on Table 1, it can be seen that the fishing boat in Indonesia are mainly fall into the category of boat without motor and outboard motors (64.42%). Meanwhile, motor boats are mostly (87.56%) in small sized too, specifically 10 GT and below (Table 2). In terms of livelihood level, number of poverty on fishing community in Indonesia is still very high. For example, a study in Garut and Indramayu shows that relatively 42.5% of fisher folk are in poor category (Kinseng 2017; Kinseng et al. 2014) Another study also shows that the number of poor fisher folk in West Sumatra are also high, which is 39% (Stanford et al. 2013). Based on the national agricultural census in 2013, the number of poor households in the sub-sector fishing in the sea is 23.79%.

In terms of social structure, fisher folk in Indonesia are not homogeneous. They can be divided into some social classes. As an example, in Indonesia, fisher folk are divided into four social class, i.e. big fisherman/capitalist, middle class fisher folk, small fisher folk, and the fisher folk laborers (Kinseng 2014). Basically, this category is similar with the fisherman in Canada where fisher folk in Canada are divided into small-scale (small fisher folk), intermediate-scale (middle class fisherman), large-scale (big fisherman), and fisher folk laborers (Wallace 1986). Furthermore, the fisher folk can also be separated into a certain groups based on the type of fishing gear or the type of fish caught, such as fishing rod, purse seine, tuna fishing, etc.

Various categories of fisher folk have consequences in organizing the fisher folk. The diversity of fisher folk drives diverse interests as well, which could cause conflicts among fellow fisher folk. To resolve these kind of issues, it is a necessity to unite the fisher folk in an organization. However, as Muszynski said, "It proved easier to organize shore workers than to

Table 1. Number of boats by category and size of ships in 2014.

Category and size of boat	2014 Total	%
Non powered boat	165,066	26.38
Outboard motor	238,010	38.04
Inboard motor	222,557	35.57
Total	625,633	100

Source: Center for Statistics and Information Data of the Ministry of Maritime and Fisheries Affairs 2015.

Table 2. The number of marine fishing vessels/vessels by ship size in 2014.

Size of boat	Total	%
<5 GT	153,493	68.97
5-10 GT	41,374	18.59
10-20 GT	14,301	6.43
20-30 GT	9,578	4.30
30-50 GT	1,029	0.46
50-100 GT	1,766	0.79
100-200 GT	840	0.38
>200 GT	176	0.08
Total	222,557	100

Source: Center for Statistics and Information Data of the Ministry of Maritime Affairs and Fisheries 2015.

bring all fishers into one union", to explain how diverse the interests of fisher folk in Canada (Muszynski 1986). The question is, what are the implications of this phenomenon for the forms of fishing organizations in Indonesia?

Since the end of the New Order era (or post 1998), fisher folk organizations have constantly growing in many parts of Indonesia. For example, in Pantura, Java, the organization of fishing movements includes: (1) Farmers and Fisher folk Awakening Fronts (FFAF) in Banten, (2) Nusantara Fisher folk Union Federation (NFUF) in Banten, (3) Traditional Fisheries Unions (TFU) in Indramayu, (4) Farmers and Fisher folk's Union (FFU) in Indramayu, (5) Tegal City Fisheries Association (TCFA) in Tegal, and (6) Indonesian Fisher folk and Farmers Movement (IFFM) in Batang. Meanwhile, fisher folk social movement in West Sumatra was driven by the Bagan Community and the West Sumatra Fisher folk Alliance. However, these fisher folk organizations are faced with rules or law from government that could have different perspectives in organizing fisheries sector. The livelihood of fisher folk is closely related with many aspect: nature, other fisher folk (that use other fishing gear or have different catch), sellers, government, and other related groups. In order to maintain this relationship, a social struggle is needed where fisher folk reacts to rules/laws that might disturb their fishing activities or any kind of injustices (Bavinck et al. 2018)

Wallace (1986) propose several types of fishing organizations in Canada, namely associations, union, and cooperatives. Association members tend to be independent fisher folk and act like small capitalists, fishing workers tend to join unions, while cooperatives are mainly formed when capital fails to meet market demand. For this study, we apply organization theory in studying social movements (McAdam et al. 1996). Organization is seen as a "collective vehicle" both in formal or informal way, to mobilize and to gather for a collective action.

This research was conducted on the fishing community in Pati District, Central Java Province. This research is part of research on the organization and social movements of fisher folk carried out in five locations on the north coast of Java such as: Jakarta Bay, Indramayu,

Tegal, Pati and Lamongan. Each location has unique community characteristics and organized fishing community.

Fishing organizations can have diverse backgrounds, for example due to geographical similarities, similarities in fishing gear or the basis of ship ownership. The uniqueness of the characteristics behind the formation of this fishing organization raises research questions in the form of: (1) what fishing organizations that exist in Pati District?; (2) what characteristics do each fisherman organization have in Pati District?; and (3) what role does each fishing organization in Pati District play?

2 METHODOLOGY

This research conducted in Bajomulyo Village and Bendar Village of Juwana District, Pati Regency, Central Java, from March until September 2018. Juwana District was chosen as research location because this district have the biggest fishers' community around Pati Regency. Data collection was conducted using survey methods, focused group discussion (FGD), in depth interview, and documentation at both villages. In depth interviews were conducted to key informants such as fisher folk opinion leader, fisher folk, village officials, members and activists of fisher folk organization. We have interviewed 50 fisher folk in total, using a "stratified-accidental sampling" (Kinseng et al. 2013) in choosing our informants.

Table 3. List of Participants

	Participants	
No.	Indepth Interviews	Focused Group Discussion (FGD)
1.	MDI as a fisher folk opinion leader	3 person from regional marine and fisheries services
2.	RJN as a fisher folk opinion leader	1 person from village officials
3.	HRI as a fisher folk opinion leader	5 person from fisher folk opinion leader
4.	MGI as a member and activists of fisher folk organization	5 person from members and activists of fisher folk organization
5.	SRO as a member and activists of fisher folk organization	
6.	MTH as a member of fisher folk organization	
7.	JSN as a fisher folk	

3 RESULTS AND DISCUSSIONS

The geographic condition of Bajomulyo Village and Bendar Village makes the residents choose fisheries as their main source of income. These villages are located on side of Silugonggo river that directly flows to Java Sea. Related offices to fisheries sector are also located in these villages (such as Fish Auction Place) which makes it easier for fisher folk to load or unload their fishing necessities and to sell their catch.

There are two type of fisher folk organization in Pati Regency: a grass root organization and national-based organization. There is one national-based organization that opened its local office and seven active grass root fisher folk organizations in Juwana. The national-based organization is called as Himpunan Nelayan Seluruh Indonesia/Association of Fisher folk in Indonesia (HNSI) for Pati Regency. This organization is the representative of HNSI from Pati Regency which is located in Bajomulyo Village. This organization was formed and organized with a hierarchical structure. The general structure of this organization are: Board of Directors in National Level, Board of Directors in Provincial Level, Board of Directors in District Level.

HNSI is a legal organization which established in 1972 under a political party called Golongan Karya (Golkar) affiliation. However, this organization has detached themselves at 1980, to maintain their ideology and neutrality as a fisher folk organization. This organization then have some branches in coastal area, especially in fisher neighborhood. HNSI in Pati Regency have around twelve grassroot organization as members from all district in Pati Regency. Seven of them located in Juwana District.

This branch office of HNSI in Pati plays role as a representative of their national organization in local area. HNSI – Pati maintaining national-local relation between the central and local members. Not only maintaining the relation among fisher organization, HNSI also played role to represent fisher as the spokesman to government. They also have some social programs such as assisting local fisher to manage shipping permit.

For grassroot based organization, there were several grassroot organization that exist in both Bajomulyo and Bendar Village. Each of this organization already have an office and most of them located in Bendar Village. These grassroot organizations establish based on fishing gear. There was a purse seine fisher organization, freezer ship fisher organization, squid ship fisher organization, cantrang fisher organization, longlines fisher organization, and holler ship fisher organization.

There are seven grassroot fisher organization found in Juwana District: Paguyuban Rukun Santoso, Paguyuban Mina Santosa, Paguyuban Mina Wahyu Sejati, Paguyuban Bangkit Nelayan Tradisional, Paguyuban Kapal Cumi, Paguyuban Rukun Sejahtera and Paguyuban Mina Melati. After this seven grassroot organizations, Paguyuban Mina Melati is the organization for fisher's wives. This fisher's wives organization conducted some post-harvest business and also money lending activities.

One of the biggest grassroot fisher organization in Juwana District is Paguyuban Mina Santosa. This paguyuban is cantrang based organization, with more or less 173 ships member. This organization established with social function such as fisher aid. This aid was for unfortunate fisher who had accident in fishing activities. Paguyuban Mina Santosa conduct some business activities such as renting fish basket (for moving fish from the ship to the TPI), and administration sevices (shipping permit services). This organization also collecting retribution from its member per trip. This retribution will be managed to fisher aid, social aid and also for supporting fisher social movement.

Lately, this organization has become the strongest grass root organization in Indonesia. Government, especially Ministry of Fisheries become the common enemy of many fisher folk organizations since the cantrang ban policy in 2015. Ministry of Fisheries released Peraturan Menteri Kementerian Perikanan Number 02/2015 and Peraturan Menteri Perikanan Number 71/2006 regarding the ban of cantrang gear for all fisher folk in Indonesia. This policy drew a big disappointment from cantrang fisher folk to the ministry.

Development of cantrang ban policy started from 1980. There was a policy about elimination of trawl nets which included cantrang. In 1997, Ministry of Fisheries stated that cantrang fishing gear is excluded as trawl nets, and were only used for small fisher folk. This statement changed in 2010, when the Ministry of Fisheries identified cantrang as trawl and banned, according 1980 policy. In 2011 until 2014, cantrang operation was allowed only in particular size of pocket and have to operate above 4 miles with maximum vessel size under 30 GTs. Recently, after 2015 and so on, cantrang fishing gear is prohibited from operated, the transition period for fisherman to switch their fishing gear to another legal fishing gear is given until July 2017.

Paguyuban Mina Santosa have been acting as the leader of cantrang fisher folk social movement. They did several demonstrations both in provincial and national level in early 2018 in front of President Palace in Jakarta. Paguyuban Mina Santosa leaders also had some meetings with The Fisheries Minister, Susi Pudjiastuti and even Joko Widodo, President of Republic Indonesia. Their effort to communicate with government officials showed a good result, where government decided to delay the ban of cantrang. President stated that they give a longer transition period for fisher folk that change their fishing gear from cantrang to other fishing gear. The reason why fisher folk struggle for cantrang legislation is based on their statement: '…cantrang struggle is all about livelihood. We (through paguyuban) will always fight for it'.

Both type of organizations, whether the grassroots or national based have their own role in Juwana fisher folk community. Durkheim said there are two types of social solidarity. first is mechanical solidarity which emphasized the social characteristic similarity between people that drives them to conduct social cooperation. The second is organic solidarity, which is based on functional differences due to the division of labor or specialization. The differences between individuals make them collaborate to complement each other's role (Thijssen 2012). Based on Durkehim explanatation, the national organization based tend to have an organic solidarity and in the other side, the grass root organization tend to play mechanical solidarity among members.

Some informants stated that paguyuban (in this case Paguyuban Mina Santosa) have stronger ties with fisher folk than HNSI. Local Paguyuban also have more simple organizational structure than HNSI. National based organization have more formal approach and formal relationship with another stakeholders such as government and other fisher folk organization.

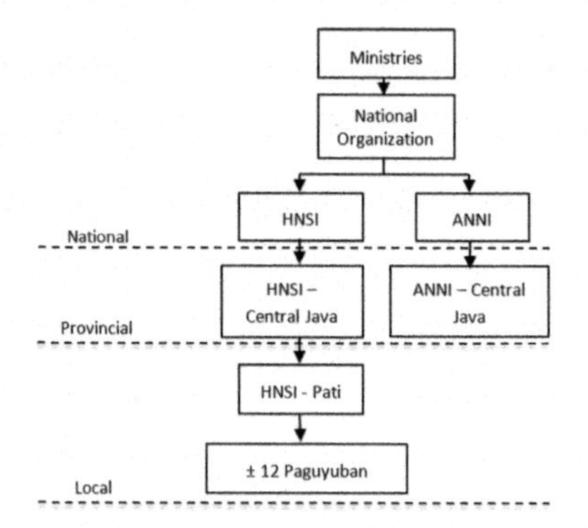

Figure 2. Fisher folk organizations network in Juwana 2018.

Grassroots organization that created by the fisher locally have more fluid relation. They maintain less formal organization (the Paguyuban itself have similar meaning with group). They also have organizational structure, but the vice chairman showed up as the real leader of this organization. The vice chairman took care of almost all administration, politics and also have more capability in person in maintaining social network between their organization with another stakeholder and organization around them. The members often gathered in *angkruk* (local terminology for informal gathering place near the river bank). This less formal approach and the similar experience as cantrang fisher folk has created stronger bond between the members. Their shared problems about cantrang ban policy also strengthen their sense of belonging to this organization.

Another unique characteristic of fisher folk organization is their membership. Only the owner of the ship and the captain written as member of fisher organization. This treatment applied for both national and grassroot based organization. In fisher folk community, there is an unique stratification which divided into two stratifications: 'in land' stratification and 'in ship' stratification. In land, there are ship owner at the top of level, ship managers (a middle man, that take care of administration of the ship and also bridging the owner with the captain) at the second layer, and the ship crew. Terminology of 'ship crew' in in land stratification refer to all ship members, start from the captain to the real ship crew. In ship stratification have several layers start from the captain at the top layer, co-captain in the second place, boatswain, chef, and the ship crew. As one of ship's crew said: '...organization is captain's business. We just followed the captain'.

This structure can be identify as structure because each layer consist of different rights, responsibility and also talks about access. The ship owner as the highest layer have the main power and rights after the ship. Including all benefit and profit. The captain have the main power in decision making on sailing activities. The ship crew tend to follow their Captain's order and only have limited access in decision making.

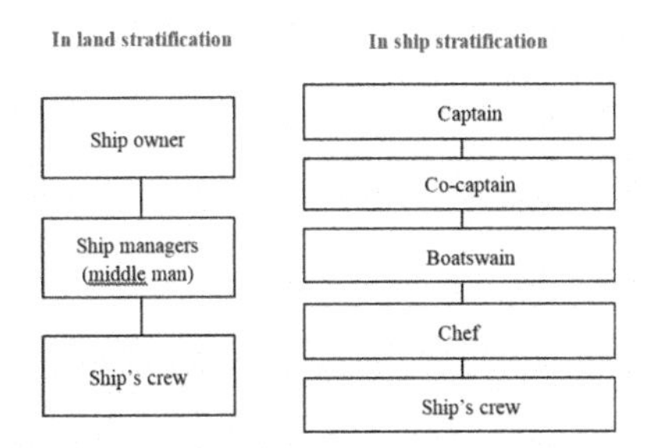

Figure 3. Fisher folk stratification in Juwana 2018.

The reason why membership of fisher organization only consist of ship owner and the captain is about the ties between worker and the ship. The ship owner and captain usually always staying serving the same ship for long time than the other member of the ship's crew. Both of the owner and the captain responsible to take care of their ship's crew and this responsibility made they counted as the formal representative of the ship. They become the spokesman for the whole ship. As stated by the ship's crew: 'a ship's crew can serve a ship and move to other ship easily. But the captain remain in the same ship'.

General role of fisher organization in Juwana District is divided in three functions: economic function, social function, and politic function. In economic, each organizations supported themselves by playing role as a dummy form of cooperative. They accept retributions from their members and return it to their members in other form such as membership aid, parcels, and other forms. They also conduct some business by offering some services such as administration services and tools renting services. Profit from this business is used to maintain the organization's daily expenses. Fisher folk organization also played role as business hub. They connect the members with another business organization such as with shipping tools vendors, engine seller, and the others. As stated by the vice chairman of fisher folk organizations: '..we collected retribution from each ship. We also provide administration services for fisher folk. That's how we maintain this organization'.

As social organization, fisher organization became the hub for all their members. Organization also provide aid for each members, not only financial aid, but also legal support in case their members have some legal problems with the cop or military army in the sea. Fisher folk organization (mostly the grass root organization) also provide aid for broader communities and village development, such us involved in building some public facilities and streets. Fisher folk organization also became the information hub between the fisher folk. They provide formal to less formal information such as updates on government policies and also internal information for their member. This function can bridge the fisher folk with government. Fisher folk organization also conduct some community development. In this case, Paguyuban Mina Melati became the most suitable example for doing community development by their programs with fisher folk wives. As stated by one member of fisher folk organization: 'paguyuban was for fisher aid, in case there was an accident. But now paguyuban also helps the village development and also for fisher folk's struggle'.

Fisher folk organization also have politic function. Both of the national and grass root based organization, facilitate the fisher folk social movement. They have been supporting and advocating fisher folk struggle against cantrang ban policy, fishing territory policy, sailing permission and other fisheries issues. Fisher folk organization also has been advocating fisher law protection to protect their members in the sea. They also played role as network hub by representing their members to other fisher organizations abroad.

4 CONCLUSION

Fisher folk organization play important roles in fisheries communities. There are about thirteen fisher folk organization identified as an active fisher folk organizations in Juwana

District. Those organizations divided in two types: national based fisher folk organization and grass root based organizations, which included one national based organization: Himpunan Nelayan Seluruh Indonesia (HNSI) and twelve grassroot organization.

National based organization is established with a top down approach as the branch of national organization. This kind of organization shows formal and strict pattern in their management. Grass root organization established by the local fisher folk with a bottom up approach, made them have stronger bond than the national based organization. Only ship owner and captain became the member of organization, based on their ties to the ship. Owner and the captain considered as their ship representatives as the hold the responsibilities after all ship crews.

Fisher folk organization plays economic, social and political role in Juwana fisher folk community. Fisher folk organization become informational and business hub for all the members and also collect retributions which will be returned to the members and broader community in both villages. Fisher folk organizations also has been facilitating fisher folk social movement to fight laws or rules that they believe do not in favor with Juwana fisher folk needs.

ACKNOWLEDGEMENT

This research is under Minister of Research, Technology and Higher Education of Republic Indonesia fund in Scheme: Penelitian Dasar Unggulan Perguruan Tinggi 2018. This research can be done by lot of help from Fisher Organization and Movement in Indonesia research's Team. Greetings and appreciation is given to all fishers in Bajomulyo and Bendar Village, seniors fisher as the elder representatives of Bajomulyo and Bendar fisher, both Village Government, and Pati Marine and Fishers Department.

REFERENCES

Arisandi. 2016. Inkosistensi Kebijakan Penggunaan Jaring Trawl (Studi Kasus Penggunaan Jaring Trawl oleh Nelayan Wilayah Perairan Gresik). *JKMP* 4(1):1-18.

Bavinck, M., Jentoft, S. & Scholtens, J., 2018. Fisheries as social struggle: A reinvigorated social science research agenda. *Marine Policy.*

Kinseng, R.A., 2017. Decentralisation and the Living Conditions and Struggle of Fishers: A Study in West Java and East Kalimantan. *Journal of Sustainable Development* 10(2):

Kinseng, R.A., 2014. *Fisher folk Conflict*, Jakarta: Yayasan Pustaka Obor Indonesia.

Kinseng, R.A., Sjaf, S. & Sihaloho, M., 2014. Class, Income, and Class Consciousness. *Journal of Rural Indonesia* 2(1):94–104.

Kinseng, R.A., Sjaf, S. & Sihaloho, M., 2013. Class, Income, and Class Consciousness. In H. Darwis, R. Muhammad, & S. Simmau, eds. *Indonesia's Maritime Community: Challenges, Opportunity, and Development*. Makassar: Department of Sociology, Hassanudin University.

McAdam, D., McCarthy, J.D. & Zald, M.N., 1996. *Comparative Perspectives on Social Movements: Political Opportunities, Mobilizing Structures, and Cultural Framings*, Cambridge: Cambridge University Press.

Stanford, R.J. et al., 2013. Exploring fisheries dependency and its relationship to poverty: A case study of West Sumatra, Indonesia. *Ocean and Coastal Management* 84:140–152.

Wallace, C., 1986. *The struggle to organize: resistance in Canada's fishery*, Toronto: McClelland and Stewart.

Ministry of Maritime and Fisheries Affairs. 2015. Maritime and Fisheries in numbers year 2015. *Center for Statistics and Information Data of the Ministry of Maritime Affairs and Fisheries.* Retrieved from http://sidatik.kkp.go.id/files/src/b74413c54e10ed63e28e4ae5cbdf6fa5.pdf at 11 December 2018.

Muszynski A. 1986. Class Formation and Class Consciousness: The Making of Shoreworkers in the BC Fishing Industry. *Studies in Political Economy 20*(Summer 1986):85-116.

Saleha Q. 2013. Kajian Struktur Sosial Dalam Masyarakat Nelayan di Pesisir Kota Balikpapan. *Buletin PSP* 21(1):67-75.

Syafrizal. 2018. Kebijakan Pengawasan Operasional Pukat Hela dan Pukat Tarik. Delivered at coordination meeting with Juwana Fisherman, July 19th 2018.

Thijssen, P. 2012. From mechanical to organic solidarity, and back: with Honneth Beyond Durkheim. *European Journal of Social Theory* 15(4):454-470.

Rural Socio-Economic Transformation – Kinseng et al. (Eds)
© 2019 Taylor & Francis Group, London, ISBN 978-0-367-23603-8

Urban and rural poor household food security and food coping strategy: A case in South Sulawesi

C.R. Ashari
Institut Kesehatan Indonesia, Jakarta, Indonesia

A. Khomsan & Y.F. Baliwati
Department of Community Nutrition, Faculty of Human Ecology, Bogor Agricultural University, Bogor, West Java, Indonesia

ABSTRACT: Nutritional problems arise due to food problems at the household level. An effort that can be made to prevent this case is by food-coping-strategies. This study used cross-sectional design. Results showed that the house-hold food security was at 27.1% in urban and at 32.9% in rural. The household food security in urban and rural was different. Coping behavior level-1 was performed by most urban households (30.6%), received the rice subsidy for the poor (Raskin) coupons, while rural households (45.9%) were raising chickens. Coping behavior level-2 was shown by most urban (44.7%) and rural (23.5%) households, borrowed money from their close-relatives. Coping behavior level-3 was performed by most urban (22.4%) and rural (23.5%) households that migrated to other cities/islands. Conclusion, food security in rural is better than in urban. Most of households that are food-secure and food-insecure both perform coping behaviors level 2, as well as on household food insecure.

1 INTRODUCTIONS

Food security, referring to the Food Act No.18 of 2012 in Indonesia, is defined as the fulfillment of food for the country to individuals, which is reflected in the availability of sufficient food for sustainable manner in number and quality, safe, diverse, nutritious, equitable and affordable and not contrary to religion, people's beliefs and culture, to be able to live healthy, active and productive lives. The Indonesian Government Regulation No.17 of 2015 concerning food security and nutrition has added nutritional status as an outcome of food security and nutrition.

According to Supariasa et al. (2012), nutritional problems arise due to food security problem at the household level that was the household's affordability to obtain food for all members of their household. This reflects the lack of accessibility of households to obtain food, one of which is caused by poverty. Suhardjo (1994) stated that household food insecurity is caused by poverty and low income.

Based on 2015 national food security and vulnerability map, one of the main characteristics that causes a high vulnerability to food insecurity in Indonesia is the high number of people living below the poverty line (DKP 2015). Since 2013, the number of the poor has always increased every year. In 2015, the number of the poor increased by 310,000 from 2014 and 520,000 from 2013 to 28.59 million or at 11.22% in March 2015 (Statistics Indonesia 2015b). In Indonesia, one of the indicators applied in the early detection of food insecurity through the Food and Nutrition Alert System is poor household.

Most households in South Sulawesi are included in the food vulnerable category, which is at 45.15% (Lantarsih et al. 2011). However, food security achieved at the regional level does not necessarily guarantee food security at a lower level. This happens due to the un-even distribution of household capacity in terms of food access and availability in the household

(Kartika & Ririn 2013). For this reason, it is important to measure food security at the household level (Saliem et al. 2002).

The SDT report (Total Diet Study 2014) shows that in South Sulawesi Province, the average energy intake in urban and rural areas in the group of 0-59 months of age is sufficient for AKE (Energy Adequacy Rate) (1,130 kcal) which is 1,131 kcal, while in other age groups from the age of 5-12 years until the elderly occur in deficits. This occurs because the affordability of food is very dependent on income. The poor households have a low access to food. The number of the poor in South Sulawesi Province in 2014 was 863,300 people or at 9.54% of the total population (Statistics Indonesia 2015a).

The existing differences based on aspects of the physical environment, social environment, life value, the tendency to diversify staple foods between urban and rural areas will certainly give a difference in the amount of income and the food consumption by household. The household's amount of income will affect food and non-food consumption by households in both areas. This will have an impact on the level of household food security. An effort that can be made by the community in overcoming these problems is by food-coping strategy. If the food-coping strategy is implemented, food insecurity will not occur sustainably. Food-coping strategies are positively related to the intensity of food insecurity (Ahmed et al. 2015, Ivers & Cullen 2011, Tanziha et al. 2010, Weiser et al. 2007, Kendall et al. 1996). Therefore, it is necessary to conduct a study to analyse food security of poor households in urban and rural areas and identify food-coping strategies in urban and rural households in order to achieve and maintain food security. The research questions of this study are what is the level of food security for urban and rural households? and how is the food coping strategies for urban and rural households in order to achieve and maintain their food security?

2 METHODS

2.1 *Design, place and time of research*

This quantitative study was carried out by survey study. The study took place in Makassar City, Tamalate Subdistrict, Mangasa Subdistrict, which represented the city and Sidenreng Rappang District, Maritengngae Subdistrict, Tanete Village and Takkalasi Village, which represented rural areas. Data were collected through interviews and in-depth interviews on several subjects in April-September 2016. A sample of 170 households were obtained using purposive sampling technique.

2.2 *Type of data*

The types of data collected in this study consisted of primary and secondary data. Primary data in this study were obtained using a questionnaire instrument. Primary data included the level of household food security that was measured by Maxwell's method and food-coping strategy.

2.3 *Data analysis*

The collected data were analyzed using Microsoft Excel 2010 computer program to record the database and Statistical Program Social Sciences (SPSS) version 21.0 for Windows to analyze the relation of each variable.

Data on household food security levels are data on the level of household food security obtained using the Maxwell method from Jonsson and Toole (1991) which was later adopted and developed by Maxwell et al. (2000). Maxwell's method combines two indicators of food security, namely food expenditure levels and energy consumption per adult equivalent unit (energy consumption of all household members is equivalent to an adult). This method reflects economic access (household food expenditure level) and household food consumption. Household food expenditure is the monthly cost spent on food by a household compared to the total expenditure per month.

A modification was made in this study. Modification is done on cut off consumption energy. Determination of cut off at the Maxwell method set by Jonsson and Toole (1991) and Maxwell et al. (2000) is 80% which is quite high if applied in Indonesia. Therefore, modifications are made in accordance with the conditions in Indonesia according to the Regulation of Minister of Agriculture No. 65 of 2010 governing minimum service standards on food security at provincial and district/city levels, and Food Security Council 2009 namely households are categorized into food insecure if their energy consumption is less than 70% of the required energy sufficiency.

Table 1. Level of household food security.

	Share of Food Expenditure Level	
Energy consumption per Adult equivalent unit	Low (≤60% of total expenditure)	High (>60% of total expenditure)
Enough (>70& energy sufficiency)	Food secure (1st category)	Vulnerable (2nd category)
Less (≤70% energy sufficiency)	Lack of consumption (3rd category)	Food Insecure (4th category)

Food expenditure level formula:

$$\text{Level of food expenditure} = \frac{\text{Household food expenditure}}{\text{Total household expenditure}} \quad (1)$$

Household food consumption is obtained through interviews using 2 x 24 hour food recall. Energy consumption per adult equivalent unit is obtained by the following formula:

$$\text{KE} = \frac{\text{KErt}}{\text{JUED}} \quad (2)$$

Where KE=energy consumption per adult equivalent, KErt=real energy consumption of households and JUED=Number of equivalent units of adults (One equivalent unit of an adult is equivalent to a man aged 30-49 years with a weight of about 62 kg and AKE of 2625 kcal, meaning that age member family below and above age that is equalized with a man aged 30-49 years).

$$\text{JUED} = \frac{\sum \text{AKErt}}{2625} \quad (3)$$

Where ΣAKErt = amount of household energy sufficiency.

$$\text{JUED} = \frac{\sum \text{AKErt}}{2625} \times 100\% \quad (4)$$

Where %KE=percent of energy consumption per adult equivalent and TE=energy consumption per adult equivalent.

Food coping strategy data were collected through interviews by asking respondents the 29 lists of coping behaviors which were then grouped into 7, that are (1) increasing in-come; (2) changes in eating habits; (3) additional immediate access to food; (4) the addition of immediate access to buy food; (5) changes in distribution and frequency of meals; (6) having a day without eating; and (7) drastic steps. Of these seven groups, it was divided into three levels, namely level 1 (increasing income, changing eating habits and accelerating access to food), level 2 (accelerating access to cash purchases, changes in distribution and frequency of meals, passing days without food), and levels 3 (making drastic steps). Level 1 consists of 13 questions, level 2 consists of 12 questions, and level 3 consists of 4 questions.

3 RESULTS AND DISCUSSIONS

3.1 *Household food security*

The proportion of households based on the level of food security was measured using the Maxwell method is shown in Table 2.

Table 4 shows that most urban households are included in the Food Insecurity category, while some rural households are included in the Food Vulnerable category. Chi-square test results shows that there is a difference in the food security level measured using the Maxwell method between urban and rural households.

In this study, the level of food security of rural households is better than the level of food security of urban households (Rachman & Supriyati 2010). This is related to the good

Table 2. The proportion of households based on the level of food security measured using the Maxwell method.

	Urban area		Rural Area		
Level of food security	n	%	n	%	p*
Food secure	23	27.1	28	32.9	0,000
Vulnerable	10	11.8	30	35.3	0,000
Lack of consumption	19	22.4	12	14.1	0,000
Food insecure	33	38.8	15	17.6	0,000

* *Chi-Square* test, p <0.05

Table 3. The proportion of households that perform coping behaviors based on 1, 2 and 3 levels.

		Urban		Rural area		Total	
No	Coping Behavior	n	%	n	%	n	%
Coping Behavior Level 1							
1.	Seeking for a side job	5	5.9	15	17.6	20	11.8
2.	Plant crops that can be eaten in the garden/land near the house	0	0	33	38.8	33	19.4
3.	Raising chickens, etc.	0	0	39	45.9	39	22.9
4.	Buying cheaper food	12	14.1	7	8.2	19	11.2
5.	Reducing the type of food consumed	14	16.5	9	10.6	23	13.5
6.	Change the priority of food purchases	1	1.2	1	1.2	2	1.2
7.	Reducing food portions	2	2.4	0	0	2	1.2
8.	Receive food from family	11	12.9	33	38.8	44	25.9
9.	Receive raskin coupons	26	30.6	28	32.9	54	31.8
Coping Behavior Level 2							
1.	Take savings	1	1.2	0	0	1	0.6
2.	Pawning assets	2	2.4	1	1.2	3	1.8
3.	Selling unproductive assets	0	0	4	4.7	4	2.4
4.	Selling productive assets	0	0	6	7.1	6	3.5
5.	Borrowing money from close relatives	38	44.7	26	30.6	64	37.6
6.	Borrowing money from distant relatives	12	14.1	10	11.8	22	12.9
7.	Borrowing money from pawnshops	1	1.2	0	0	1	0.6
8.	Buying food with debt in a shop	17	20.0	3	3.5	17	10.0
9.	Changes in food distribution	1	1.2	0	0	1	0.6
10.	Reducing the frequency of meals per day	1	1.2	0	0	1	0.6
Coping Behavior Level 3							
1.	Migration to other cities/villages/islands	19	22.4	20	23.5	39	22.9

Table 4. Proportion of households doing behavior coping based on the level of food security.

No.	Coping Behavior	Food resistant n	Food resistant %	Not food resistant n	Not food resistant %	p*
1.	Not do behavior coping	0	0	0	0	
2.	Level 1	13	25.5	25	21.0	
3.	Level 2 and Level 1 and 2	22	43.1	68	57.1	0.228
4.	Level 3, Level 1 and 3, Level 2 and 3, Level 1,2,3	16	31.4	26	21.8	

* *Chi-Square* test, $p < 0.05$

adequacy of food in rural household (Patel et al. 2015, January 2014, Rachman & Supriyati 2010, Sukandar et al. 2006).

Rural areas are identical to agricultural areas that can be planted with staple food crops, such as rice, cassava, and corn that make a major contribution to consumption, which in turn it will affect household food security (Ibok et al. 2014). In addition, most households in rural areas still have land around the house/garden. This significantly affects household food diversity (Fry et al. 2015), which in turn it will affect the level of household food security. The land around the house/yard in this study was utilized by households to plant food crops such as cassava and vegetables, and to raise livestock such as chickens.

Rahmawati et al. (2014) and Musotsi et al. (2008) also suggested that farming can improve household food security. This is because farming can provide direct access to food that can be picked and consumed by household members every day, thus providing food sources of vegetables and fruits that are rich in vitamins and minerals (Musotsi et al. 2008). The more number of plants and livestock a household has, then the food supply in the household can increase. This will have an impact on the level of household food security. Livestock and agricultural crops have a mutually beneficial relationship.Livestock get food from grass or agricultural crop waste, and vice versa, livestock manure can be used as manure that can fertilize agricultural land and increase agricultural crop production (Musotsi et al. 2008).

3.2 *Food coping strategies*

According to Maxwell et al. (1999), food-coping strategy is an effort made to overcome unfavorable conditions including the decreasing access to food. A person can work by relying on intellectual abilities, physical abilities, and material. According to Maxwell (1995), a list of food-coping strategies carried out by households to meet their needs of food, among others 1) eating foods that are less preferred; 2) limiting the portion of food; 3) borrowing food or money to buy food; 4) mothers limit their food portions; 5) reducing the type of food; 6) having days without food; 7) drastic steps. Furthermore, Usfar (2002) divided the seven groups into 3 levels, from the least effort in providing food (level 1) to extreme measures that can have a negative impact on households (level 3).

Coping behavior at level 1 was carried out by most urban households who received *Raskin* coupons. This occurred because the households as respondents were poor households based on National Population and Family Planning Board (*BKKBN*) criteria and they were recipients of *Raskin* coupons. Whereas in rural areas, most households raised chickens. This is because the research location in rural area is one of the chicken breeders in Sidenreng Rappang District. This is in line with the research by Rahmawati et al. (2014) stating that most of the research subjects maintain livestock.

The type of livestock most widely kept is chicken because chicken breeding is relatively easy and inexpensive to maintain compared to other livestocks. Chicken can also be sold or consumed immediately if needed. Coping behaviors level 1 that were not performed by urban and rural households in this study were buying food with lower value, collecting wild food, food for work from the government, and food barter.

The coping behavior level 2 carried out by most of urban and rural households was by borrowing money from close relatives because the loan process and repayment are easy, still based on a family principle in accordance with the deal of both parties. According to Mutiara et al. (2008), the tendency to borrow money is done by close relatives because of the trust of the siblings, hence they do not feel reluctant and ashamed to borrow when they need it. This is in line with statement by Chagomoka et al. (2016), Connolly-Boutin & Smit (2016), Walsh & Rooyen (2015), Hadley et al. (2012), Tanziha et al. (2010), and Coutsoudis et al. (2000) that a coping strategy that can be done if income is insufficient is by borrowing money from relatives or neighbors. The coping behaviors level 2 that were not done by the urban and rural households in this study were borrowing money from loan sharks and passing days without food.

Coping behavior level 3 carried out by urban and rural households was migration to other cities/villages/islands. This is in line with the research by Warner & Afifi (2014), Rademacher-Schulz et al. (2014) and Quaye (2008) who migrated to find a better living. However, in this study, the households that migrated had not succeeded. Some respondents in urban areas were migrants, who came to the city to work, but they remained in the category of poor families because of odd jobs and irregular income which was related to the low education level. Similar to rural households, the household members went abroad to get job opportunities, but their lives were still difficult in overseas because odd jobs were associated with low education, thus many of them could not transfer money/income to their families in the village, but rather asking for money for living overseas. In this study, migration has not succeeded in overcoming economic problems among urban and rural households. Coping behaviors at level 3 which were not performed by urban and rural households in this study were becoming migrant workers, giving children to relatives, and families divorced.

Table 4 presents the proportion of households that perform coping behaviors based on the food security level that was measured using the modified Maxwell method. Table 4 shows that the majority of households included in the food-resistant group and those in the non-food-resistant group performed coping behaviors level 2. Chi-square test results indicated that there was no difference between the coping behavior levels in the food-resistant group and the non-food-resistant group. Both of them performed coping behaviors level 2.

4 CONCLUSIONS

The rural household's food security level is better than the urban household's.Coping behavior level 1 performed by most of urban households is receiving the rice subsidy for the poor (*Raskin*) coupons, while the coping behavior level 1 performed by most of rural households is raising chickens. The coping behavior level 2 performed by most of urban and rural households is borrowing money from close relatives. The coping behavior level 3 performed by most of urban and rural households is migrating to other cities/villages/islands. Most of households that are food secure and food insecure both perform coping behaviors level 2.

Based on this study, it was found that the level of food security in rural households was better than urban households, so the community specially in urban needed to be provided with counselling and/or training from the government or related institutions regarding methods of maintaining food security by empowering communities such as consuming cheaper local food, gardening or growing local food and raising chickens, and others. So, their coping strategy is more varied which is expected to achieve and maintain their food security.

4.1 *Suggestions*

Research on the non-poor households needs to be conducted to determine the food security status. Communication on nuclear family program (family planning) needs to be improved.

ACKNOWLEDGEMENT

The author would like to thank the Makassar City Government and the Sidenreng Rappang District Government for assisting in the implementation of this research.

REFERENCES

Ahmed, U.I. Ying, L. & Bashir M.K. 2015. Food insecurity and coping strategies by micro growers in punjab, Pakistan. *Journal of Environmental and Agricultural Sciences* 3: 31-34.

[BPS] Badan Pusat Statistik. 2015a. *Makassar dalam Angka 2015*. Makassar: BPS Kota Makassar.

[BPS]. 2015b. *Indikator Kesejahteraan Rakyat 2015*. Jakarta: BPS Jakarta.

Chagomoka, T. Unger, S. Drescher, A. Glaser, R. Marschner, B. & Schlesinger, J. 2016. Food coping strategies in northern Ghana. A socio-spatial analysis along the urban–rural continuum. *Agricultural & Food Secure* 5(4): 1-18.

Connolly-Boutin, L. & Smit, B. 2016. Climate change, food security, and livelihoods in sub-Saharan Africa. *Regional Environmental Change* 16: 385–399.

Coutsoudis, A. Maunder, E.M.W. Ross, F. Ntuli, S. Talor, M. Marcus, T. Dladla, A.N. & Coovadia, A.M. 2000. *South Africa: A Qualitative Study on Food Security and Caring Patterns of Vulnarable Children in South Africa*. Geneva: World Health Organization, Nutrition for Health and Development, Sustainable Development and Healthy Environments.

[DKP] Dewan Ketahanan Pangan, Kementerian Pertanian dan World Food Programme (WFP). 2015. *Peta Ketahanan dan Kerentanan Pangan Indonesia 2015*. Jakarta: Dewan Ketahanan Pangan, Kementerian Pertanian dan World Food Programme (WFP).

Fry, H.H. Azad, K. Kuddus, A. Shaha, S. Nahar, B. Hossen M. Younes, L. Costello, A. & Fottrell, E. 2015. Socio-economic determinants of household food security and women's dietary diversity in rural Bangladesh: a cross-sectional study. *Journal of Health, Population and Nutrition* 33(2): 2072-1315.

Hadley, C. Stevenson, E.G.J. Tadesse, Y. & Belachew, T. 2012. Rapidly rising food prices and the experience of food insecurity in urban Ethiopia: Impacts on health and well-being. *Social Science & Medicine* 75(12): 2412-2419.

Ibok, W.O., Idiong I.C. Brown, I.N. Okon, I.E. & Okon, U.E. 2014. Analysis of food insecurity status of urban food crop farming households in cross river state, Nigeria: A USDA Approach. *Journal of Agricultural Science* 6(2): 132-141.

Ivers, LC, Cullen, K.A. 2011. Food insecurity: special considerations for women. *The American Journal of Clinical Nutrition* 94(6): 1740–1744.

January, I. 2014. Tingkat ketahanan pangan rumah tangga petani dan pengaruh kebijakan raskin. *Jurnal Ekonomi Pembangunan* 15(2): 109-116.

Jonsson, U. & Toole, D. 1991. *Household Food Security and Nutrition: A Conceptual Analysis*. New York: UNICEF Mimeo.

Kementerian Kesehatan, R.I. 2014. *Buku Survei Konsumsi Makanan Individu dalam Studi Diet Total (SDT) 2014*. Jakarta (ID): Badan Penelitian dan Pengembangan Kesehatan.

Kendall, A. Olson, C.M. & Frongillo, E.A. 1996. Relationship of hunger and food insecurity to food availability and consumption. *Journal of the American Dietetic Association* 96(10): 19–24.

Lantarsih, R. Widodo, S. Darwanto, D.H. Lestari, S.B. & Paramita, S. 2011. Sistem Ketahanan Pangan Nasional: Kontribusi ketersediaan dan konsumsi energi serta optimalisasi distrubusi beras. *Analisis Kebijakan Pertanian* 9(1): 33-51.

Maxwell, D. Klemeser, M.A. Rull, M. Morris, S. & Aliadeke, C. 2000. Urban livelihoods and food nutition security in greater Accra, Ghana. IFPRI in collaborative with Noguchi Memorial for Medical Research and World Health Organization. *Research Report No 112*.

Maxwell, D. Clement, A. Levin, C. Margaret, A. Sawudatu, Z. & Grace, M.L. 1999. Alternative Food Security Indicators: Revisiting the Frequency and Severity of Coping Strategies. *Food Policy* 24(4): 411-429.

Maxwell, D. 1995. *Measuring Food Insecurity: The Frequency And Severity Of "Coping Strategies"*. Washington DC: Food Consumption and Nutrition Division International Food Policy Research Institute 1200 Seventeenth Street, N.W.

Musotsi, A.A. Sigot, A.J. & Onyango, M.O.A. 2008. The role of home gardening in household food security in butere division of western Kenya. *African Journal of Food Agriculture Nutrition and Development* 8(4): 375-390.

Mutiara, E. Sjarief, H. Tanziha, I. & Sukandar, D. 2008. Analisis strategy food coping keluarga dan penentuan indikator kelaparan. *Media Gizi dan Keluarga* 32(1): 21-31.

Patel, K. Gartaula, H. Johnson, D. & Karthikeyan, M. 2015. The Interplay between Household Food Security and Wellbeing Among Small-Scale Farmers in the Context of Rapid Agrarian Change in India. *Agriculture &Food Security* 4:16.
Quaye, W. 2008. Food security situation in northern Ghana, coping strategies and related constraints. *African Journal of Agricultural Research* 3(5): 334-342.
Rachman, H.P.S. & Supriyati. 2010. Konsumsi protein hewani dan peningkatan kualitas sumber daya manusia di Provinsi Nusa Tenggara Barat. *PANGAN* 20(1): 8192.
Rademacher-Schulz, C. Schraven, B. & Mahama, E.S. 2014. Time matters: shifting seasonal migration in Northern Ghana in response to rainfall variability and food insecurity. *Climate and Development* 6(1): 46-52.
Rahmawati, W. Erliana, U.D. Habibie, I.Y. & Harti, L.B. 2014. Ketahanan pangan keluarga balita pasca letusan Gunung Bromo, Kabupaten Probolinggo, Indonesia. *Indonesian Journal of Human Nutrition.* 1(1): 35-49.
Saliem, H. Lokollo, E. Purwantini, T.H. Ariani, M. & Marisa, Y. 2002. Analisis ketahanan pangan tingkat rumah tangga dan regional. Buletin Agro Ekonomi. Volume 2. Sari MR, Prishardoyo. 2009. Faktor-faktor yang memengaruhi kerawanan pangan rumah tangga miskin di Desa Wiru Kecamatan Bringin Kabupaten Semarang. *JEJAK* 2(2): 135-143.
Suhardjo. 1994. *Pengertian dan Kerangka Pikir Ketahanan Pangan Rumah Tangga.* Bogor: PSKPG, LP, IPB.
Sukandar, D. Khomsan, A. Riyadi, H. Anwar, F. & Eddy, S. 2006. Studi ketahanan pangan rumah tangga miskin dan tidak miskin. *Gizi Indonesia* 29(1): 22-32.
Supariasa, Bachyar, B. & Ibnu, F. 2012. *Penilaian Status Gizi.* Jakarta: EGC.
Tanziha, Hardinsyah, & Ariani, M. 2010. Determinan intensitas kerawanan pangan serta hubungannya dengan food coping strategies dan tingkat kecukupan energi di kecamatan rawan dan tahan pangan. *Jurnal Gizi dan Pangan* 5(1): 39–348.
Walsh, C.M. & Rooyen, F.C.V. 2015. Household food security and hunger in rural and urban communities in the free state province, South Africa. *Ecology of Food and Nutrition* 54: 118–137.
Warner, K. & Afifi, T. 2014. Where the rain falls: Evidence from 8 countries on how vulnerable households use migration to manage the risk of rainfall variability and food insecurity. *Climate and Development* 6(1): 1-17.
Weiser, S.D. Leiter, K. Bangsberg, D.R. Butler, L.M. Percy-de Korte, F. Hlanze, Z. Phaladze, N. Iacopino, V. & Heisler, M. 2007. Food insufficiency is associated with high risk sexual behavior among women in Botswana and Swaziland. *Plos Medicine* 4(10): 1589–1597.

Rural Socio-Economic Transformation – Kinseng et al. (Eds)
© 2019 Taylor & Francis Group, London, ISBN 978-0-367-23603-8

Strategy for improving women's leadership capacity in Baran Village, Central Java Province, Indonesia

N. Purnaningsih & A. Fatchiya
Department of Communication and Community Development Science, Faculty of Human Ecology, Bogor Agricultural University, Bogor, West Java, Indonesia

Y. Saraswati & A. Wibowo
Centre for Gender and Child Studies, Bogor Agrictulural University, Bogor, West Java, Indonesia

ABSTRACT: Empowering women is one of the targets of Sustainable Development Goals (SDGs). In Permendesa No. 5/2015, priority for the use of village funds has been arranged, so the proposal of women leaders in development has not been accommodated and realized. This paper intends to explain the capacity of women leaders and the strategy of increasing their leadership capacity. The survey was conducted on all female leaders in Baran Village with 18 leaders and 5 informants. The role of female leaders is high in education programs, health, savings and loans, and recitation. The role of male leaders in the village institutional system dominates the village Musrenbang process, BPD Institutions, LPM, and RT-RW Leaders. The strategy for improving women's leadership capacity is directed at the type of transformational leadership, in which the relationship between leaders and followers is based on shared values and beliefs. Indicators of transformational leadership include capabilities in terms of: focus, communication, caring, creating opportunities, and credibility.

1 INTRODUCTION

There has been a change in attitudes and roles of women, that women do not only act as housewives but also have active roles in the public sector. At the same time women carry out domestic sector work and reproductive functions, while playing a role in the public sector both socially, economically and politically. However, there are some people who argue that women are less suitable to be leaders, because men are considered more capable (Putra 2009).

Although many women have become leaders like heads of village, women's rights to choose and to be elected are still injustice. This is because of patriarchal culture. Humaidah (2012) states that because of patriarchal culture, the opportunities for women to participate in public were obstructed by perception of women's role and position itself. This is confirmed by Usman (2012), states that the representation of women in government of Aceh Province is occured at low level of bureaucracy. There were 38 percent of women working at staff position, 26 percent at Echelon IV, 11 percent at Echelon III, and 4 percent at Echelon II. This means that women are still recognized as lacking in their abilities, because they are feminine so that they are considered unable to carry out government tasks (Mahmudah 2011). The involvement of women in the legislature, acting in the political world is still relatively small, including in countries where the level of democracy and human rights equality is quite high (Karim 2004).

Leadership is the ability to influence the behavior of other people in certain situations in order to be willing to work together to achieve a predetermined goal (Putra 2009). The Tao Te Ching in Shaskhin & Shaskhin (2011) states that the best relationship between leaders and followers is to serve one another. The meaning of mutual service relationships comes from how much this leader has influence. Shaskhin & Shaskhin (2011) divide the

Table 1. Leadership type matrix based on the influence process and leader-subordinate relationship pattern.

Indicator	Leadership Type		
	Charismatic leadership	Transactional leadership	Transformational leadership
Influence Process	Compliance through identification, with the expectation of being in power like a leader	Willingness to expect reward (or avoid punishment)	Internalization of values that become guidelines for acting together
Motives of Leadership Power	Controlling other people	Controlling with other people	Empowered subordinates are guided by a shared vision
Motives of Subordinate Power	Subordinates have dependency	People who achieve achievements independently	Interdependent, empowered subordinates - as work partners

three types of leadership based on the process of influence and pattern of leader-subordinate relationships. These types of leadership consist of charismatic, transactional, and transformational leadership types. In more detail see the differences in the three types presented in Table 1.

Shaskhin & Shaskhin (2011) describe efforts towards transformational leadership. In influencing the members, a leader has several ways to lead ideal leadership. Shaskhin & Shaskhin mention the leader's actions in transformational behavior must have the following things: (a) focus; the actions of leaders must be able to grab people's attention and focus them on important issues in a discussion; (b) communicative; leaders must have communication skills which mainly are listening effectively, prioritizing appropriate feedback, and explaining complex ideas clearly; (c) credibility; has consistent behaviors (reliable), especially in fulfilling promises; (d) respective leadership (care); giving actions that show concern for subordinates, such as expressing congratulations on achievements made by subordinates; (e) risk leadership (creating opportunities); actions are designed to get the full commitment of each individual to new ideas and projects, and most of the time, leaders must be able to manage human resources by involving them and giving them clear responsibilities.

Every woman has chance to become a leader. However, Candraningrum (2012) states that the biggest challenges in women's leadership are 'untrained' women in public decision-making acivities, and also the perspective of seeing women considered as second-class citizen. Another obstacle that excluded women from decision-making processes is, the lack of access to knowledge and information combined with the social isolation (Janssens 2010). According to Fitriani (2015), found that women tends to has transformational leadership, which characterized by democratic and participative perspective. Related to this statement, women also have the basics charecteristics of a leader, which are: empathy, multitasking, have many network, and good negotiator (Fitriani 2015).

In order to influence women's leadership, there are several factors which are: (a) individual characteristics including age, education level, occupation, income level, attitude and motivation to lead; (b) Environmental characteristics: program management and opportunities, group involvement, media accessibility, and interaction with outside parties. Women's leadership isn't influenced by sex differences, but more intended for individual characteristics and job completion itself (Situmorang 2011).

This paper intends to analyze how the women's capacity is in the context of rural development. Rural women have participated significantly in various fields of development, therefore the research questions are: (1) What is women's leadership capacity (2) What is the strategy for increasing women's leadership capacity. This paper intends to explain the condition of women leaders' capacity and the strategy for increasing the leadership capacity of women leaders.

2 METHODS

This study uses a quantitative approach with survey research methods and is supported by qualitative data. This research was conducted in Baran Village, Sukoharjo District, Sukoharjo Regency, Central Java. The study was conducted on November, 2015. The unit of analysis in this study was female individuals who were formal leaders and opinion leaders (activist cadres, entrepreneurs, female leaders, group leaders). The survey method is collecting data using a questionnaire on selected samples. Survey method with saturated samples are carried out on all members of the population called the census. This method is often done if the population is relatively small, less than 30 people, or research that wants to make generalizations with very small errors (Sugiyono 2012). In this research, census was conducted on all female leaders in Baran Village (18 people). Census method technique is used to select respondents. Discussions with 5 informants included village heads, community leaders and village officials.

Questionnaire data was analyzed descriptively. Data was processed statistically using SPSS software (Statistical Package for Social Sciences) for Windows version 21.0. The statistical test used was the Wilcoxon different rating test. This test is used to analyze whether there are significant differences from the Y variable in the 2 characteristic categories of variable X. For example, leaders who are young and old (variable X) are different in terms of their leadership abilities (Variable Y).

3 RESULT AND DISCUSSION

3.1 *Internal characteristics, environmental characteristics, and its effect on women's leadership*

Table 2 presents data on internal characteristics of female leaders in Baran Village. Young age categories (15–40 years) and old age (41–70 years). Occupation level refers to the division of the level of occupation that is perceived to be of high and low value, that are considered high are civil servants, military personnel, private sector, entrepreneurs, while the non-working population, farmers, are considered low.

Table 2. Internal characteristics of women leader at Baran Village (n=18).

Internal	High		Low	
characteristics	amount	percent	amount	percent
Age	13	72.2	5	27.8
Occupation level	4	22.2	14	77.8
Education level	14	77.8	4	22.2
Income level	9	50.0	9	50.0
Leadership attitude	15	83.3	3	16.7
Motivation to lead	9	50.0	9	50.0

Higher education level includes Diploma, Bachelor, Master, Doctoral, whereas, while not in school – Senior High School is considered low. High income is more than 2,000,000 IDR. The attitude of leadership is considered high if you like to lead while considered low if you don't like being a leader. Motivation to lead is high if you are on your own and low if there is family encouragement, or you are invited by others, and get rewarded.

The majority of female leaders in this study belong to the old age group (72.22%). The old age group has work activities in the village, while the younger age group mostly works outside the village. The majority of female leaders are old age groups who work as housewives so their work values are considered low.

As explained by Mrs. SWN (48 years old, one of the leaders in the Education Sector):

...in this village the young people are out of the village, some are in college and some are working. Those who are here live as mothers from the older generation. But it does not rule out the possibility that we always cadre and give young people the opportunity to study for the betterment of the village ...

Female leader's income level is 50 percent high. Most female leaders (83.3%) stated that they wanted to become leaders, and participated in building villages.

Table 3 presents data on the environmental characteristics of female leaders in Baran Village. Involvement of female leaders in deciding programs at meetings and Musrenbang is 50 percent. This means that women leaders are very given the opportunity to attend, hold opinions, organize activities, lead activities, etc. and are given the opportunity to participate in decision making in meetings and Musrenbang. The involvement of female leaders in group activities is low, this means that the involvement of women leaders in planning, program implementation, program evaluation involving group members is still low. The level of media accessibility for female leaders tends to be low. The interaction of women leaders with outside parties of female leaders tends to be low. The frequency of using the media by female leaders is still low, including internet, TV, radio, newspapers, magazines, banners and brochures, so that media literacy is considered low. The interaction of women leaders with the village parties and parties outside the sub-village, district, other groups that have the same program, and related experts in each activity implementation, is still low.

Table 4 presents data on the results of the Wilcoxon Difference Test between internal characteristic variables with the ability to influence. All internal characteristic variables relate to the real influence ability for female leaders, except the income variable.

Table 4 in the form of a summary of the results of the Wilcoxon test shows that the higher the level of individual characteristic variables, the higher the leadership ability, which are the level of focus, communicative, credible, caring and creating various opportunities; except income level variables. Women leaders with old age have the ability to be more focused, more communicative, more caring, and more able to create opportunities than young people. All

Table 3. Environmental Characteristics Of Women Leader at Baran Village (n=18).

Environmental	High		Low	
characteristics	amount	percent	amount	percent
Program management and opportunities involved	9	50.0	9	50.0
Group involvement	8	44.4	10	55.6
Media accessibility	6	33.3	12	66.7
Interaction with outsiders	6	33.3	12	66.7

Table 4. The results of the wilcoxon difference test between internal characteristic variables with the ability to influence (leadership).

Individual	the ability to influence (leadership)				
characteristics	focus	communicative	credible	caring	creating opportunities
Age	√	√	√	√	√
Occupation level	√	√	√	√	√
Education level	√	√	√	√	√
Income level	x	x	x	√	x
Leadership attitude	√	√	√	√	√
Motivation to lead	√	√	√	√	√

√: significantly correlated to alpha 0.005, x not significantly correlated to alpha 0.05

Table 5. The results of the wilcoxon difference test between enviromental characteristic variables with the ability to influence (leadership).

Environmental	the ability to influence (leadership)				
characteristics	focus	communicative	credible	caring	creating opportunities
Program management and opportunities involved	√	√	√	√	√
Group involvement	√	√	√	√	√(-)
Media accessibility	√(-)	√(-)	√(-)	√(-)	√(-)
Interaction with outsiders	√(-)	√	√(-)	√(-)	√(-)

internal variables influence leadership abilities: focus, communicative, caring, and creating opportunities, except income level variables. Variable income level influences caring. Rich people are more capable of caring, in this case contributing a portion of their income to development in the community.

Table 5 presents data on the results of the Wilcoxon Difference Test between environmental characteristic variables with the ability to influence. All environmental characteristic variables relate to the real influence ability for female leaders. The higher the level of environmental characteristic variables the higher the leadership ability, which are the level of focus, communicative, credible, caring and creating various opportunities except income level variables.

Power and leadership have inseparable relationships. Individual power in leading members becomes the individual's main capital to influence its members. French & Raven (1960) conveys the types of power seen from the source or origin, which are reward power, legitimate power, coercive power, referent power and expert power. Table 6 shows the number and percentage of female leaders based on the choice of the type of power used to lead. Leaders who use power in the form of rewards (praise and self-development opportunities) have 33 percent. Leaders who apply legitimate power or power based on formal or structural positions are 72 percent. No leader applies coersive power or coercion, pressure, threats and punishment. Leaders who apply referent power, which is the power that comes from charismatic self, by being a role model, are 72 percent. Leaders who apply expert power or power derived from their abilities and professionalism are amount 56 percent. Following are the reasons why coersive power was not seen in the power system of female leaders in Baran Village:

> *... Because the organization whose purpose is to build this village, is based more on sincerity and dedication. We have never implemented a coercive system. Those who disobey the rules in the community will feel social sanctions on their own. In leading we need to set a good example and the effort in increasing our capacity is enough to influence members ...*
>
> *(MJ, 64)*

Table 6. Amount and percentage of women leaders using power in Baran Village (n=18).

Type of power	amount	percent
Legitimate	13	72.0
Referent	13	72.0
Expert	10	56.0
Reward	6	33.0
Coercive	0	00.0

Amount of power used in influencing
1 type of power (33 percent); 2 types of power (44 percent)
3 types of power (6 percent); 4 types of power (17 percent)

In the process of influencing, a leader often uses more than one type of power. The most used types of power are referent power and legitimate power.

3.2 *Type of leadership based on leader-follower relations*

The type of leadership is identified from 3 indicators, which are the characteristics of follower compliance, how leaders influence, and how the attitude of citizens towards leaders. The attributes of these indicators are as follows in Table 7, showing percentage of Female Leaders based on the Type of Leadership Indicator Executed.

Table 7 presents the percentage of female leaders based on indicators of the type of leadership that is carried out. Based on these indicators, it was concluded that the type of female leadership carried out was more characterized by the type of transformational leadership.

3.3 *Strategy of women's leadership capacity development: transformational leadership*

All variables are related to the real leadership capacity of women in influencing the community, except income. Internal variables include: age, work, level of education, attitude and motivation. Environmental variables include: program management, group involvement, media accessibility, and interaction with outside parties. Leadership capacity includes: focus, communicative, caring, and credible.

The capacity of women's leadership in this case is in the implementation of development programs in general, not only in the priority of the use of village funds. Figure 1 presents a diagram of relationship between factors that influence leadership. The leadership capacity of women is directed at the type of transformational leadership, meaning that leadership is based

Table 7. Percentage of female leaders based on the leadership type indicator conducted in Baran Village (n=18).

No.	Indicator	Charismatic leadership	Transactional leadership	Transformational leadership
1	Characteristics of follower compliance	100	39	100
2	How leaders influence	50	94	100
3	The attitude of followers	72	94	72
	Average score	74.0	75.7	90.7

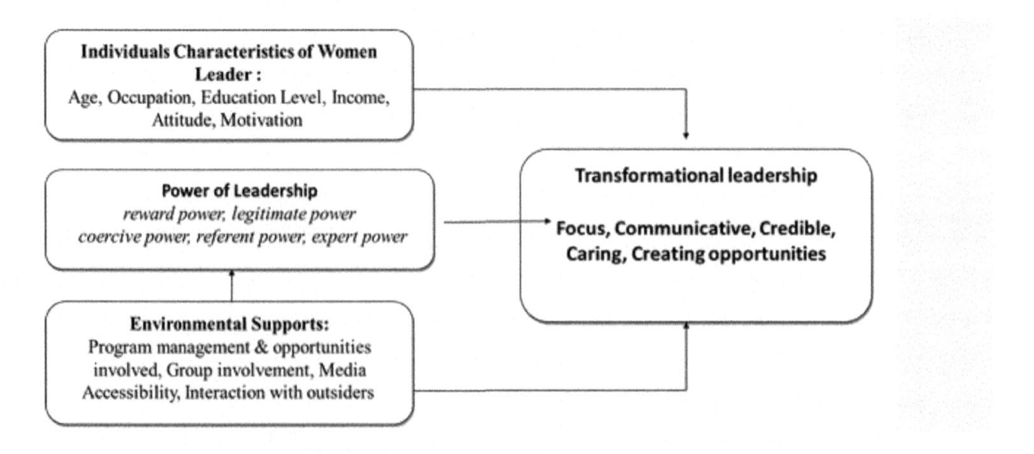

Figure 1. Strategy of women's leadership capacity development: transformational leadership

on shared values and beliefs/visions between leaders and subordinates, not on personal characteristics or leader figures and not on transactional material delivery.

3.3.1 *Internal capacity building for women's leadership*

1. Age of leaders is generally and relatively old. In the future it is expected that younger leaders will emerge. The average age at this time is 46 years. Leadership training for youth, and providing youth opportunities for development activities in the village, will increase the capacity of young leaders. The flow of youth urbanization to outside or inside the village.
2. Job leaders usually provide regular income every month, such as civil servants. Being a leader in a village is voluntary work that does not generate income. In this case it needs to be encouraged for those who have permanent jobs to be involved in development activities, contribute to development in the village.
3. Educational leaders are usually secondary and tertiary education graduates. Those who are highly educated usually have jobs and have a steady income. There are several cases of female leaders who are highly educated even though they do not work outside the home, and claim to be housewives. In this case, it is necessary to encourage highly educated women to be involved in village development programs.
4. The attitude of women leaders in this case is a positive attitude and supportive for women leaders in various fields of development. Most of the time, this attitude prevents women themselves from becoming leaders, unwillingness, feeling inadequate, feeling not in the field of women, etc. which is a barrier to women's participation.
5. Motivation of women to be leaders will be strong if the motivation comes from themselves; their own desires. Encouragement from outside parties, such as the family, the invitation of the village head, other people, or because of other rewards or awards is another reason women become leaders. In general in the village, respect for leaders is usually in the form of recognition from the community, recognized as a leader. Being involved in organizations also makes women have an identity as female leaders, therefore attributes such as uniforms are important, because they are symbols of identity and social status.

3.3.2 *Development of leader power*

Leader power is the ability of leaders to influence others. In this study, power is grouped into 5 types, based on the origin or source of power, namely: reward power, legitimate power, coercive power, referent power and expert power. These types of power can be obtained due to innate nature and is learned and collected by individual leaders. Power can also be influenced by internal characteristics such as age, level of education, income, attitudes and motivations of leaders even though the effects vary or do not always arise in all leaders.

Leader power is accumulative, meaning that the more sources of power possessed, the more powerful and influential the leader will be. To achieve a transformational type of leadership, a leader needs to have several sources of power based on his function in leadership, logically will be explained in the following section.

1. Referent Power; a power that comes from references that are owned, such as positive personal traits, so that the better the traits possessed by a leader, the more people (subordinates) will like. This reference power is power derived from charisma, individual characteristics, being role model or having attractive personalities, such as honest, fair, intelligent, persistent, hardworking, kind, compassionate, generous, etc. Which are attractive characteristics of a leader who becomes power in influencing. Can this referent power be learned or obtained as one of the leaders' strengths? The answer is yes. A good level of education, (formal and non-formal) interactions with other parties (people inside and outside the village), a good environment (affection from people around), etc. are considered able to affect the growth of charismatic traits, although often mentioned that charismatic characteristics are derived from heredity.
2. Legitimate Power comes from someone's formal position, for example, the structural position in the organization is in the form of formal authority to use and control organizational resources. This position of authority in power includes reward power and coercive

power as well as in its application. Phenomena found in everyday life often occur that an informal leader does criticism, also wants to move the community but does not have legitimate power or no position, so that the influence cannot be effectively implemented. Occupying a certain position is one form of strength for an informal leader so that his influence is recognized by society, or his leadership is formally recognized.

3. Expert Power comes from one's ability and professionalism. The process of influencing other people occurs because of the expertise possessed by the leader. This power can be influenced by the level of formal and non-formal education, even though it is possible for someone to have self-taught skills or self-study. Usually someone is appointed as a leader because of his expertise in a particular field. In the case of rural communities where the level of education is not something that is considered important, a leader must equip with other sources of strength such as referent power and legitimate power.
4. Reward Power comes from appreciation (reward) or a form of benefit from something given by the leader. Reward power is not always manifested in material forms such as salary, promotion, and position, but it can also be something psychological in nature such as praise and opportunities for self-development or learning.
5. Coercive Power is the power possessed by someone to influence others, because he has a strong position. This form of power is an order which if disobeyed, there will be threats, sanctions or even punishment. The implications of applying this power are usually in the form of pressure, fear, lack of confidence and stress from subordinates. In milder levels, this coercion can be manifested in the form of pressure, which still has positive aspects from subordinates. The situation of a depressed organization can bring new spirit, group solidarity and find creative ways to solve problems together, and in the long run can empower.

3.3.3 *Development of environmental conditions for women's leadership*

1. Program management is the management of development programs that give women the opportunity to be involved. The involvement of female leaders in development programs is expected to be facilitated by each development program, which provides opportunities for women to start from program planning, implementation and evaluation activities. This is a condition for the sustainability of village development programs in the future.
2. Group involvement for women is a condition that allows women to be involved in groups/ organizations or organizing activities. This will create opportunities for women to learn organizationally and play a role in every development activity, foster self-confidence, practice communication skills, develop a sense of solidarity, and as part of self-actualization needs.
3. Accessibility Media means the availability of communication media at the local level, both in the form of print media and electronic and digital media. Media development can be an opportunity for women leaders to be more media-literate, open to a variety of information. This is important for a leader, as an information seeker, to open up the horizons of the community, to develop development programs that are specifically local in global scrutiny. Leaders who are exposed to information are usually more broad-minded, cosmopolitan, and innovative.
4. Interaction with outside parties for a leader is a necessity for program development and management. Outside the village, especially sub-districts and districts are very influential parties in the planning and management of development programs. Developing opportunities to interact with parties outside the village is one of the leadership capabilities in building networks (social networking).

4 CONCLUSIONS

1. The position and role of female leaders have been significantly involved in development programs and village institutional systems, although their participation is still limited to around 10 to 20 percent. The role of female leaders is high in education programs (PAUD),

health (Posyandu), PKK, savings and loans, and recitation. The position and role of male leaders in the village institutional system dominates the village Musrenbang process, BPD Institutions, LPM, and RT-RW Leaders.

2. Individual characteristics: age, education level, and attitude are high, motivation to lead and income level is average, accupation level is low
3. Enviromental characteristics: group involvement, media accessibility, interaction with outsiders are low, program management & opportunities involved is average
4. The Wilcoxon test statistics test proves that there are differences in women's leadership capacity in rural areas. Internal factors, including age, occupational level, level of education, attitudes, and motivation, while environmental factors including the opportunity to participate provided by program management, involvement in groups, accessibility of the media, and opportunities to interact with parties outside the village in terms of managing development development programs real with women's leadership capacity
5. The type of leadership applied by female leaders is transformational leadership
6. The strategy of developing women's leadership capacity towards transformational leadership is through developing internal, environmental characteristics, and developing leader power. Transformational leadership is a type of leadership in which the relationship between leaders and followers is based on shared values and beliefs/vision between leaders and subordinates, not on personal characteristics or leader figures and not on transactional material giving. Indicators of transformational leadership capacity include ability in: focus, communication, caring, creating opportunities, and credibility

REFERENCES

Candraningrum, D. 2012. Kepemimpinan perempuan indonesia: tantangan dan peluang. *J. Perempuan* 17 (4): 73–84.

Fitriani, A. 2015. Gaya kepemimpinan perempuan. *J. TAPIs* 11(2): 1–24.

French, JRP & Raven, B. 1960. *The Bases of Social Power : Origins and Recent Developments.Chapter 20.* Downloaded at: http://www.communicationcache.com/uploads/1/0/8/8/10887248/the_bases_of_social_power_-_chapter_2c09_-_1959.pdf

Humaidah, L.N. 2012. Affirmative action dan dampak keterlibatan perempuan: sebuah refleksi. *J. Perempuan* 17(4): 73–84.

Janssens, W. 2010. Women's empowerment and the creation of social capital in indian village. *J. World Development* 38(7): 974–988.

Karim, A.J. 2004. Kepemimpinan Wanita Madura. *Mimbar: Jurnal Sosial dan Pembangunan* 13 (2):221–234.

Mahmudah, W.U. 2011. Kepemimpinan Kepala Desa perempuan dalam lembaga pemerintahan desa (Studi kasus di Kecamatan Sukorejo, Kabupaten Kendal). Thesis. Semarang: Departement of Sociology and Anthropology Education, Universitas Negeri Semarang.

Putra, A.C. 2009. Persepsi tentang kepemimpinan perempuan di kalangan pelajar pria SMK Negeri 6 Surakarta (kajian dari sudut pandang kesetaraan gender). Thesis. Surakarta: Faculty of Teacher Training and Education, Universitas Sebelas Maret.

Saskhin, M & Saskhin, M.G. 2011. *Prinsip-prinsip Kepemimpinan.* Surabaya: Erlangga.

Situmorang, N.Z. 2011. Gaya kepemimpinan perempuan. *Proceeding of PESAT (Psikologi, Ekonomi, Sastra, Arsitektur & Sipil)*, Universitas Gunadarma (ID): 18–19 Oktober 2011. Hal: 129–135

Sugiyono. 2012. *Metode Penelitian Kuantitatif Kualitatif dan R&D.* Bandung: Alfabeta.

Usman, N.A. 2012. Potret keterwakilan perempuan dalam pemerintahan di aceh. *J. Perempuan* 17(4): 73–84.

Rural Socio-Economic Transformation – Kinseng et al. (Eds)
© 2019 Taylor & Francis Group, London, ISBN 978-0-367-23603-8

Author Index